How to Talk to a Science Denier

How to Talk to a Science Denier: Conversations with Flat Earthers, Climate Deniers, and Others Who Defy Reason

Lee McIntyre

The MIT Press
Cambridge, Massachusetts
London, England

This book was set in Stone Serif and Stone Sans by Westchester Publishing Services. Printed and bound in the United States of America.

Library of Congress Cataloging-in-Publication Data

Names: McIntyre, Lee C., author.
Title: How to talk to a science denier: conversations with flat earthers, climate deniers, and others who defy reason / Lee McIntyre.
Description: Cambridge, Massachusetts : The MIT Press, [2021] | Includes bibliographical references and index.
Identifiers: LCCN 2020050478 | ISBN 9780262046107 (hardcover)
Subjects: LCSH: Science—Social aspects. | Science—Public opinion. | Pseudoscience. | Truthfulness and falsehood. | Reasoning.
Classification: LCC Q175.5 .M3954 2021 | DDC 306.4/5—dc23
LC record available at https://lccn.loc.gov/2020050478

10 9 8 7 6 5 4 3 2 1

For Mohamad Ezzeddine Allaf, MD
A healer

A man with a conviction is a hard man to change. Tell him you disagree and he turns away. Show him facts or figures and he questions your sources. Appeal to logic and he fails to see your point.

—Leon Festinger, *When Prophecy Fails* (1956)

It is easier to fool people than to convince them that they have been fooled.

—Mark Twain (attributed)

Contents

Introduction

I admit, I hesitated when I first put on the lanyard I'd received from a smiling young woman in a white lab coat staffing the check-in table at the Flat Earth International Conference 2018. I wondered if anyone would recognize me—was that someone taking pictures? But then again, why would they? I'd been sitting in my office studying science denial for the last fifteen years. With my flannel shirt and badge, I looked just like the rest of them. It was the "cloak of invisibility" I needed for a philosopher of science gone undercover, at least for the first twenty-four hours.

After that, I would be ready to make my move . . .

Suddenly, I felt a hand on my shoulder and turned to find a man in a black T-shirt, smiling with an outstretched hand. His shirt said "NASA LIES."

"Hey, welcome, Lee," he said. "So tell me, how did you get into Flat Earth?"

For a number of years it has been fairly clear—at least in the United States—that truth is under assault. Our fellow citizens don't seem to listen to facts anymore. Feelings outweigh evidence, and ideology is ascendant. In an earlier book, I explored the question of whether we now live in a "post-truth" era, where facts and even reality itself are up for grabs . . . and what the consequences of that might be.[1] What I found was that the roots of today's "reality denial" go straight back to the problem of "science denial," which has been festering in this country since the 1950s, when the big tobacco companies hired a public relations expert to help them figure out how to fight the science that said smoking was linked to lung cancer.[2] This scheme provided a blueprint for how to wage a successful campaign of misinformation against whatever topic one liked—evolution, vaccines, climate change—with the result that we now live in a society where two people can look at the same inauguration photograph and come to opposite conclusions about how many people were in attendance.[3]

The political mess in Washington will be with us for a while. But the fallout for science is already an emergency. A recent report from the United Nations Intergovernmental Panel on Climate Change (IPCC) warns that we have reached a dangerous tipping point.[4] The effects of global warming are happening much faster than expected, and many countries have already missed their targets for the Paris Climate Agreement. The polar ice cap could be gone by 2030; the coral reefs could disappear by 2040; sea levels in New York and Boston could rise by as much as five feet before the end of the century.[5] A few years back, UN Secretary-General António Guterres warned that "if we do not change course by 2020, we risk missing the point where we can avoid runaway climate change."[6] Meanwhile, as of this writing, the climate-denier-in-chief in the White House continues to promote the fantasy that climate scientists have a "political agenda" and that even if climate change is happening, it is not provably "man-made" and could "very well go back."[7] Unfortunately, millions agree with him.

How do we reach them? How can we get people to change their minds based on facts? It has sometimes been thought that you can't. In fact, some have said that trying to do so would lead to a backfire effect, where we'd make the problem worse by causing partisans to double down on their mistaken beliefs.[8] This led to several overheated headlines such as "This Article Won't Change Your Mind" (*Atlantic*) and "Why Facts Don't Change Our Minds" (*New Yorker*).[9] But there is a problem with this mentality, because in the last few years new research has shown that the original backfire effect could not be replicated.[10] Yes, people are stubborn and resist the idea of changing their beliefs based on facts, but for most change is *possible*. And if we don't try, things will only get worse.

In one of the most exciting recent developments, in June 2019, a landmark study was published in the journal *Nature Human Behaviour* that provided the first empirical evidence that you *can* fight back against science deniers.[11] In an elegant online experiment, two German researchers—Philipp Schmid and Cornelia Betsch—show that the worst thing you can do is *not* fight back, because then misinformation festers. The study considered two possible strategies. First, there is content rebuttal, which is when an expert presents deniers with the facts of science. Offered the right way, this can be very effective. But there is a lesser-known second strategy called technique rebuttal, which relies on the idea that there are five common reasoning errors made by *all* science deniers. And here is the shocking thing: both strategies are equally effective, and there is no additive effect, which means that *anyone* can fight back against science deniers! You don't have to be a scientist to do it. Once you have studied the mistakes that are common to their arguments—reliance on conspiracy theories, cherry-picking evidence, reliance on fake experts, setting impossible expectations for science, and using illogical reasoning—you have the secret decoder ring that will provide a universal strategy for fighting back against *all* forms of science denial.[12]

Unfortunately, there is one crucial thing that Schmid and Betsch left out. There are essentially three possible levels of engagement with science deniers: inoculation, intervention, and overturning belief. Schmid and Betsch dealt only with the first two.[13] In a sympathetic commentary that ran in the same issue of *Nature Human Behaviour*, Sander van der Linden explains that Schmid and Betsch's methodology could be useful to pre-identify the bogus techniques that science deniers use, in an attempt to "pre-bunk" them, so that their impact on a potential audience can be mitigated. Second, Schmid and Betsch demonstrate that even when participants have recently been exposed to scientific misinformation, it is effective to immediately intervene and explain the faulty reasoning, before mistaken beliefs have time to set in cement. Both pre-bunking and debunking are potentially powerful tools that are vindicated by their findings. What the researchers did *not* do, however, was *measure whether it was possible to overturn the beliefs of hard-core science deniers*, especially those who had already been exposed to years' worth of scientific misinformation. Schmid and Betsch (and van der Linden) brilliantly deal with the *audience* for science deniers . . . but what about those who were already committed science deniers before they participated in the study?

Here, unfortunately, the empirical literature leaves us adrift. Anecdotal accounts have suggested that the best way of convincing someone to change their beliefs is through direct personal engagement—but the Schmid and Betsch study was all done online. Yet doesn't it make sense that if we are trying to convince people to change their minds, it would help to build some trust first? Most beliefs are formed within a social context (and not based solely on facts), so shouldn't social context matter in changing them?

In his important essay "How to Convince Someone When Facts Fail," professional skeptic and historian of science Michael Shermer recommends the following strategy:

> From my experience, (1) keep emotions out of the exchange, (2) discuss, don't attack (no ad hominem or ad Hitlerum), (3) listen carefully and try to articulate the other position accurately, (4) show respect, (5) acknowledge that you understand why someone might hold that opinion, and (6) try to show how changing facts does not necessarily mean changing worldviews.[14]

If you listen to the stories of science deniers who have altered their beliefs, they universally report the positive influence of someone they trust. Someone who built a personal relationship with them and took their doubts seriously, then shared the evidence. Facts alone were not enough. In two recent reports on overcoming vaccine denial, former anti-vaxxers (or at least those who were vaccine-hesitant) report having had their outlook changed by people who sat down with them, listened to all of their questions, and explained the answers with ample patience and respect. During the 2019 measles outbreak in Clark County, Washington, the state government sent public health officials out to "meet with parents in small groups or one-on-one, sometimes for hours at a time, to answer their questions." As a result, one woman reported having "changed her mind, deciding to give her kids the shots after a doctor at a vaccine workshop answered her questions for more than two hours, at one point drawing diagrams on a whiteboard to explain cell interaction. He was thoughtful, factual and also 'still very warm,' she said."[15]

In another account, a South Carolina resident wrote of her own conversion on the subject of vaccines, in a *Washington Post* op-ed titled "I Used to Be Opposed to Vaccines. This Is How I Changed My Mind":

> My reasons for being against vaccines stemmed mostly from misunderstanding the ingredients in the vaccines and how they worked. People who tried to convince me not to vaccinate told me about the many ingredients in vaccines, such as aluminum salts, polysorbate 80 and formaldehyde but they did not explain the purpose of these ingredients. . . . What changed my mind? It was finding a group of people who were strongly in favor of vaccines and willing to discuss the topic with me. They were able to correct all the misinformation I had heard and respond to my concerns with credible research and other helpful information.[16]

On the topic of climate change, one encounters a similar anecdotal literature, including the remarkable account of a rock-ribbed Republican politician,

Jim Bridenstine, whom President Trump appointed as chief administrator at NASA, who changed his mind on global warming within a few weeks of taking his new job. Back in 2013, Bridenstine had given a speech on the House floor in which he falsely claimed that "global temperatures stopped rising ten years ago." He now says, "I fully believe and know that the climate is changing. I also know that we human beings are contributing to it in a major way. Carbon dioxide is a greenhouse gas. We're putting it into the atmosphere in volumes that haven't been seen, and that greenhouse gas is warming the planet. That is absolutely happening, and we are responsible for it." What changed his mind? For one thing, he says that he "read a lot." But he did so within the context of surrounding himself with new colleagues at NASA—where he "heard a lot of experts" and soon concluded that there was "no reason to doubt the science" on climate change.[17]

Respect, trust, warmth, engagement. These are the common threads that run through such first-person accounts. Schmid and Betsch offer us important experimental evidence about the best strategies for dealing with science deniers. But for whom and within what social context? Schmid and Betsch is a landmark study, but it leaves open perhaps the most intriguing question in the science denial debate: can we change the minds of even hard-core science deniers, and if so, how?

For years I have been studying the issue of science denial and trying to figure out how best to push back. I was using content rebuttal and technique rebuttal long before Schmid and Betsch vindicated them. But the problem is that—in the real world, face-to-face—one is often dealing not with the *audience* for science deniers but with the most obstinate science deniers themselves. Here, there is no question of inoculating them against misinformation, nor of intervening before it takes hold. Their beliefs have been formed by years of marinating in misinformed ideology, and often their very identity is at stake. Can *their* minds be changed too?

In my most recent book, *The Scientific Attitude: Defending Science from Denial, Fraud, and Pseudoscience* (MIT Press, 2019), I developed a theory of what is most special about science, and outlined a strategy for using this to defend science from its critics. In my view, the most special thing about science is not its logic or method but its values and practices—which are most relevant to its social context. In short, scientists keep one another honest by constantly checking their colleagues' work against the evidence and changing their minds as new evidence comes to light. But does the general public understand this? And, even if they did, how do we put that understanding into practice?

As I was out on tour to promote my earlier book, *Post-Truth*—and in anticipation of *The Scientific Attitude* (which was still in page proofs at the time)—I kept getting questions from the audience about how they could fight back. What could they say to get truth deniers to change their beliefs? My advice was to engage. To talk to people one-on-one about the scientific attitude and the importance of reason. Not to let people get away with trashing the importance of evidence as a result of their deeply misinformed views about how science worked.

Then I thought about why I myself wasn't out there doing that.

It was worth a try. Even if I could not convince any hard-core science deniers to give up their beliefs, I might at least have an effect on their audience. And if I could just channel the persuasive reasoning skills I had learned as a philosopher, maybe I could also make a dent in science deniers' claims that *they* were the ones who were actually being scientific. That they were skeptics rather than deniers. Even if I couldn't convince them with evidence, I could show where their reasoning skills weren't up to snuff. This was when I envisioned writing the book you now hold in your hands.

Thus, in November 2018, I found myself in the ballroom of the Crowne Plaza Hotel in Denver, Colorado, surrounded by six hundred shouting, clapping true believers at the Flat Earth International Conference. It felt odd to be alone in my belief that Aristarchus and Copernicus had long ago settled the question of whether the Earth was a globe. But after all those years studying science denial from my desk, I was finally here, in the belly of the beast, with perhaps the most reviled science deniers on the planet (sorry . . . in the world). Why did I start with Flat Earth? Because I wanted to choose the worst of the worst. To confront the type of science deniers that even other science deniers would make fun of.

I thought that if I could study the most elemental case of science denial, maybe I could learn how to talk to others—like climate-change skeptics— whose views might seem more moderate and nuanced. In the back of my mind I also thought that perhaps the reasoning strategies for all science deniers might be the same, and that whatever argumentative tricks I used on the Flat Earthers might work on climate change deniers too.

Little did I know what lay in store for me . . .

1 What I Learned at the Flat Earth Convention

It is unbelievable but true that Flat Earth theory is making a comeback. Although the basic science to demonstrate the curvature of the Earth is over two thousand years old—and available to any high school physics student—one now finds numerous Flat Earth meetup groups in various cities, hears their views spouted by celebrities like rapper B.o.B.[1] or NBA players Kyrie Irving[2] and Wilson Chandler, and can even attend a Flat Earth convention like the one I did—Flat Earth International Conference (FEIC 2018)—in Denver.

First, the threshold question. Are these people serious? Yes, completely so. To believe in Flat Earth is not something one would come to lightly, for Flat Earthers are routinely persecuted for their views. Many report losing jobs, being kicked out of their churches, and being ostracized from their families. Is it any wonder that many would choose to keep their beliefs private? Given this, it is nearly impossible to tell how many Flat Earthers there actually are.[3] Perhaps this accounted for the celebratory atmosphere I witnessed at FEIC 2018, where complete strangers greeted one another as old friends.

In one of the first presentations at the opening of the conference, one speaker memorably took up the mantra "I am not ashamed," for which he was greeted with wild applause. Some in the audience had tears in their eyes as they repeated the phrase to themselves: apparently they were not ashamed either. To be insulted, ridiculed, and dismissed for your views cannot be a fun experience. I think of this every time I hear someone dismiss Flat Earthers as trolls or jokesters who must be in it for the fun. Who would endure this for fun? Perhaps I am simply credulous, but in all of my time at FEIC 2018, I did not meet one person who seemed anything other than deeply committed to their beliefs. Indeed, that is probably part of what made the meeting so meaningful to its participants. Other than me and a few journalists who were there to cover the event, FEIC seemed like a revival meeting for misfits who had finally found their kin.

As I looked around the ballroom, what struck me most was that if you didn't already know what the event was about, you wouldn't be able to tell. Everyone looked so "normal." Not a tinfoil hat anywhere. Men and women, young and old, multiracial, from all walks of life.[4] I did see a lot of black T-shirts (some with funny logos), but nothing else to indicate that this was a fringe crowd. If you looked away from the three huge multimedia screens at the front of the ballroom, you might think that you were waiting for the opening act at a Metallica concert. In my casual shirt and jeans, I fit right in.

I sat toward the front, next to a couple about my age who said they were from Paradise, California. This was just a few months after the deadly wildfires there, so I asked if they were OK. The man spoke up. "Well, our house got burned up. We can't go home. And we still haven't heard from my wife's mother. She was old and had dementia. So she might be lost." This floored me. I looked discreetly at his wife, but she gave no reaction. In the middle of that situation, they had loaded up the truck and driven out to Denver for a Flat Earth convention? I expressed my sympathies, and we continued to talk about the wildfires, during which the man offered that he thought the government had been putting accelerant on the fires; he'd seen chemtrails overhead beforehand. The woman offered, "I just think there's something fishy about the fires, how they had been isolated, then closed in." Behind us sat a mother and her six- or seven-year-old son, with a spiral notebook that said "Bible Research." Then the show started.

After a rousing musical act, the opening speech was given by Robbie Davidson, the event's organizer, who talked about how he used to be a "globalist," but then got converted to Flat Earth in the process of trying to refute it. He wasn't against science, he explained, only "scientism." But "the truth shall set you free!" At that point the couple from Paradise leapt to their feet and yelled, "Praise Jesus," as the rest of the crowd erupted in applause. I just sat there, taking notes. As they sat down, the couple looked over at me. Robbie continued by pointing out—I thought largely for the media's sake—that this meeting did *not* have any affiliation with the Flat Earth Society. He went on to ridicule that group for believing that the Earth was a "flying disk" in space.[5] He implored any skeptics in the crowd that if they were going to ridicule his group, to do so with an understanding of what they actually believed. Stay for the whole conference. Do your own research. Science has had a stranglehold on our cosmological beliefs for centuries, he said, "but the foundation is crumbling!" And the crowd went crazy again.

Not all of the events were talks. In addition to the rapper who had warmed up the crowd, there was a video from Flat Earth Man, a rock star wannabe whom everyone in the crowd already seemed to know. His video "Space Is Fake" was well received (and well done), and featured all sorts of goofy Photoshopped images that were apparently shown to reinforce the message that if he could fake pictures, so could the government. NASA was the butt of most of the jokes. Here I learned that virtually all Flat Earthers believe that *all* of the pictures of Earth from space are fake, that we never landed on the Moon, and that all of the employees at NASA—along with millions of others—are "in on the conspiracy" to cover up God's truth that the Earth is flat. Those who were not already Flat Earthers were either part of the cover-up or they were sheep. To drive the point home, the video indicated that if you count up the place of the letters in the alphabet for "National Aeronautics and Space Administration," it comes to 666.

After the video I asked the man from Paradise to explain who is behind all this. He knew that I was a newbie, so maybe my cover wasn't completely blown yet.[6] He said, "The adversary." I pressed him: "The devil." He went on to explain that the devil helps those who are in power, and that includes all world leaders: every head of state, astronauts, scientists, teachers, airline pilots, and many others who are rewarded by the devil for keeping the secret of Flat Earth.[7] He then explained, "This all goes back to the Bible." There couldn't have been a flood in the time of Noah if the Earth was round, he opined.[8]

Over the next forty-eight hours I heard similar things from many other people, which were largely a combination of nonsensical physics mixed with Christian fundamentalism.[9] What impressed me, though, was that even though most of the participants seemed to have deeply held religious views, they did *not* base their belief in Flat Earth on faith. Instead, they maintained that their beliefs were based on *evidence*, both in favor of Flat Earth and against the "globalist" hypothesis. They encouraged participants to do their own experiments.[10] Indeed, the whole point of the conference, Robbie had said, was to present material for "educational" purposes. "Don't believe things based solely on authority" was a common refrain. In fact, several speakers encouraged the crowd not to believe what *they* were saying just because they had said it, but to use it as a jumping-off point to do their own research.

This apparently is how many Flat Earthers become converted. More than once, I heard someone say that they used to believe in the global Earth—for which the unkind word (that we were encouraged not to use) was "globetard"—and had tried to refute Flat Earth *but could not*, and so concluded that it must be true. "Be careful, we used to be like you," one speaker warned. In the process of trying to prove that Flat Earth was a fraud—usually after watching a series of YouTube videos—many had instead convinced themselves that it must be true. Indeed, insofar as the Flat Earthers had a method, this seemed to be it: if you can't prove that the Earth is round, then you should believe it is flat.[11] And it didn't seem to bother them one bit that most of their "research" came from watching online videos. Indeed, according to Asheley Landrum, a psychologist at Texas Tech who has studied Flat Earthers, YouTube is the gateway for virtually *all* new recruits of Flat Earth.[12]

To a person, Flat Earthers have a profound distrust in authority—and great belief in first-person sensory experience. And their standard of belief is *proof*. In their epistemology, to question a belief is sufficient for concluding that it must be false. But what about their own beliefs? In a group as skeptical as Flat Earthers, it is curious that they do not apply any real scrutiny to the basis for their *own* beliefs. If one asks *them* for proof that the Earth is flat, they normally turn the burden of proof back on the globalist. The choice is binary. If you can't prove that Earth is round—subject to their paranoid suspicions of bias or fraud about any evidence you offer—then it must be flat.

It is also curious that for a belief system which purports to be based on evidence and experiment, most Flat Earthers describe their conversion as one of revelation. One day they woke up and realized that there was a worldwide conspiracy of people who had been lying to them. Once they were willing to question the depth of the cover-up, Flat Earth was at the bottom of the rabbit hole. "Trust your eyes" became their mantra. "Water is level." "Space is fake." "A government that could lie to you about 9/11 and the Moon landing is one that would lie to you about Flat Earth." The Flat Earthers all describe their conversion as a quasi-mystical experience, where one day they "took the red pill" (and, yes, they adore the movie *The Matrix*) and realized a truth that the rest of us have been blind to for our entire lives, as a result of our miseducation and indoctrination: the Earth is flat.

What does this mean? What do they actually believe? Not only that the Earth is flat, but that the continent of Antarctica is not really a continent at all but an ice wall spread out along the perimeter of the Earth (which is what keeps the water from falling off) and that the whole thing is covered by a transparent dome, outside of which the Sun, Moon, planets, and stars (which are very close) shine through. Of course, this means that all space travel is faked (for how could they get through the dome?). And it means that the Earth does not revolve or rotate (for if it did, wouldn't you feel it?).

To state this immediately raises a series of questions:

What does this mean for gravity, the constellations, time zones, eclipses? And just what the heck is *under* the Flat Earth, anyway? Flat Earthers love these sorts of questions and have an answer for every one of them—though they sometimes vary from person to person, which is what the conference was all about.[13]

Who could keep such a secret? The government, NASA, airline pilots, and others.

Who put them up to that? "The adversary" (the devil), who rewards them mightily for covering up God's truth.

Why don't others realize the truth? Because they have been fooled.

What is the benefit of believing in Flat Earth? Because it's the truth! And it's consistent with the Bible.

What about the scientific proofs of a round Earth? They are all flawed, which is what the rest of the conference was about.

To spend two days attending seminars with titles such as "Globebusters," "Flat Earth with the Scientific Method," "Flat Earth Activism," "NASA and Other Space Lies," "14+ Ways the Bible Says Flat Earth," and "Talking to Your Family and Friends about Flat Earth" is in some ways to spend two days in an asylum. The arguments were absurd yet intricate and not easily run to ground, especially if one buys into the Flat Earther's insistence on immediate first-person sensory proof. And the social reinforcement that participants seemed to feel in finally being with their own was palpable. Psychologists have long known that there is a social aspect to belief; FEIC 2018 was a lab experiment in tribal reasoning.

The next presentation was by one of the Flat Earth superstars—Rob Skiba—whose talk had been billed as one of the main "scientific" presentations. I could hardly wait. At the beginning, Skiba pointed out that he had no academic credentials . . . but he *did* have a white lab coat, which gave him all the credibility he said he needed. He then began a lecture that included a ten-point slideshow on the "evidence" for Flat Earth (which consisted mostly of "evidence" against the global Earth). Foucault's pendulum? A fake! If it's real, then why do they need a drive motor to keep the pendulum moving? (Physics says friction.) Photos from space? He said that they were all illustrated or painted by NASA (in the era before Photoshop). During the talk I also learned that Skiba had an alternative theory of gravity (which I couldn't reproduce here if I tried), thought that Flat Earth was supported by pillars that had been put there by God (resting on what, he didn't say), and he didn't understand how water could adhere to "a spinning ball." Just try spinning a beach ball and throw a cup of water at it, and see what happens! Oh boy. What he *did* believe in was a video he showed of an elderly woman pushing a nine-ton boulder with one hand. If that was possible, he said, they must have already figured out anti-gravity. And if that was true, they could fake a Moon landing in a warehouse.

By this point my head was spinning; none of it made any sense. But then he switched to something I vaguely remembered from physics: the Coriolis effect. Skiba wanted to know why it was that if you shot a bullet east to west you had to make an adjustment, but not when you shot it north to south. Wouldn't the alleged "sideways" motion of the Earth come into play? And if it didn't, did that mean the Earth wasn't spinning? None of this comported with anything I remembered about the Coriolis effect (and I confess I didn't remember the technical details well enough to know where his description of the phenomena was at odds with reality), but what I did notice was that Skiba really didn't seem to understand inertial frames of reference. He apparently thought that if you tossed a baseball in the air on a train going at constant speed it would land behind you rather than in your mitt. Was this what he was saying about the bullet?

I was still pondering this conundrum (and wishing I remembered more physics) when the talk moved on to something I recalled very clearly from college astronomy. Skiba displayed a photograph of the skyline of the city of Chicago taken from sixty miles out in Lake Michigan.[14] This caught my attention because I remembered a lecture that had talked about the phenomenon called hull down, which is where a ship disappears on the horizon hull first, due to the curvature of the Earth. It had been a long time since freshman year, but I checked the calculation provided on screen and it was right: at sixty miles the top of the Sears Tower should have already dipped below the horizon. Indeed, you didn't have to go out that far . . . you only had to go out forty-five miles! But here was a picture of the full, shimmering Chicago skyline from sixty miles out. Proof? Well, in a group of skeptics, did it ever occur to anyone that perhaps the picture might be faked? We had just heard that virtually every single picture from NASA was fake, so why not this one?

Later, after the presentation, I caught up with Skiba by one of the booths at the swag fair of Flat Earth merchandise for sale in the adjacent ballroom.[15] There were Flat Earth maps and T-shirts, hats and jewelry. I bought a CD of Flat Earth music—which was surprisingly catchy and well done—and some stickers and a necklace for my wife. At first, Skiba must have thought I was a fan when I approached him and said that I'd just seen his presentation and had a few questions.

As it turned out, the photo was not a fake. It was a real image that demanded an explanation. During his presentation, Skiba had dismissed the correct scientific explanation for the photograph, which is due to something called the superior mirage effect. This occurs when there is a blanket of cold air (for instance, on the surface of the water) just underneath a blanket of hotter air above it. As the light travels through these layers, it is bent, as though by a lens, and an observer might see an image hovering in the air where it should not be.[16] There is nothing mysterious about this. Those who have driven on hot pavement and seen "puddles" on the surface of the road (which vanish as one approaches) have seen the inferior mirage effect, which occurs when the surface of the road is *hotter* than the air above it. In that case, the image is *below* where one would expect it; with the superior mirage effect, the image appears *above* its actual position. It is an illusion, but it is not "fake." It is a real image that one can photograph. In just the right conditions, one can even take a video of the blinking lights of a city over the curvature of the Earth's horizon. It is a cool effect.

When I asked Skiba about the superior mirage effect, he dismissed it.

"I dealt with that in my talk," he said. "It's made up."

"You didn't deal with it in your talk," I said. "You just said you didn't believe it."

"Well, I don't," he said.

We talked a bit more about the photo, and he explained that he wasn't just taking this on authority. He himself had gone out on Lake Michigan and recreated the effect from forty-six miles away. He said he'd seen it with his own eyes.

By this time, a crowd of admirers had gathered to ask Skiba their own questions, and the "scientist" was getting antsy. He'd probably figured out by then that I wasn't a Flat Earther, but he couldn't very well break away now without looking small in front of his fans.

But I had another question.

"So why didn't you go out one hundred miles?" I asked.

"What?"

"A hundred miles. If you'd gone out that far, not only the city would've disappeared but also the superior mirage too. If it didn't, you'd have your proof."

He shook his head. "We couldn't get the captain of the boat to go out that far."

Now it was my turn to scoff. "What? You've devoted your entire life to this work and you didn't go? You had the definitive experiment within reach and you couldn't go out an extra fifty-five miles?"

He turned his head and began to talk to someone else.

Looking back, maybe I don't blame him. I was too hot. Too confrontational. It's hard to stay cool when your beliefs are being challenged. Maybe I was proof of that myself.

Over the next forty-eight hours, I had numerous other less heated conversations about the Flat Earther's "evidence." Given their belief that the Earth is flat, that the continent of Antarctica as we know it does not exist, that there is a giant dome over the top of the Earth, and that the Earth does not move, there should have been ample opportunity to test the accuracy of the Flat Earthers' hypothesis. Yet in two days of talking about Foucault's pendulum, shadows during an eclipse, the International Space Station, the fact that water is subject to gravitational pull, and other matters from college astronomy, not once did I seem to disturb any of the Flat Earthers' beliefs that they were right.

The temptation to come up with some definitive experiment or scientific finding that blows Flat Earth right of the water is overwhelming at an event like this. I wanted to debunk them so badly I could taste it. But if the goal is to get a Flat Earther to admit that they are wrong, it probably cannot be done, at least not in this way. The evidence for a global Earth has been around since Pythagoras (who argued that if the Moon was round, the Earth must be also). Since Aristotle (who said that if we walked north to south we would see different stars). And since Eratosthenes (who calculated the circumference of the Earth by measuring the Sun's shadow on two sticks placed very far apart).[17] This evidence had been around for 2,300 years and the Flat Earthers already knew it, but they remained unconvinced. They had an excuse for everything.[18] So if they weren't already convinced by two millennia of physics, why would they be convinced by me?

I needed to reset.

I wasn't a physicist, and I hadn't come to FEIC 2018 to talk to them about the scientific evidence for or against Flat Earth anyway. I was a philosopher, and I had come to talk about how they were *reasoning*. The frustrating thing with Flat Earthers is that even if you find a flaw in one of their arguments or experiments, they will just look at you and say, "Yes, but what about . . ." and move on to the next thing. They have hundreds of "points" and unless you are willing to play whack-a-mole and knock down every single one of them, they will not admit defeat. For them, there is no such thing as a "definitive experiment." If they tell you that they *know* Flat Earth is true because of X, and you then show them that X is not true, they will just move on to the next thing.

This is decidedly not what scientists do. In my earlier book, *The Scientific Attitude*, I had argued that the primary thing that separates science from nonscience is that scientists embrace an attitude of willingness to change their hypothesis if it does not fit with the evidence.[19] This is reinforced not just through the commitment of individual scientists, but in the community standards of science as a whole, where they test one another's work and hold it up to the highest level of scrutiny. Is that even close to what the Flat Earthers were doing?

To be fair, some Flat Earthers had thrown down the gauntlet and said that they were willing to change their minds if presented with the right evidence. At FEIC 2018, I had the pleasure of meeting "Mad" Mike Hughes, who was famous for going up in a homemade rocket to try to see the curvature of the Earth. He didn't get very far. In his first attempt, he went up only 1,875 feet, which is shorter than the 2,717-foot-tall Burj Khalifa skyscraper in Dubai. Rather than building a rocket, he could have taken an elevator. And, without a 60-degree or wider field of vision, the curvature of the Earth is not visible until one gets above 40,000 feet. No amount of observation below that height would suffice for settling the question of the curvature of the Earth. Even if Hughes went up as high as 30,000 feet, he'd have to settle for the view he could get on most commercial aircraft.[20]

When I met Hughes, standing next to his rocket at FEIC 2018, I admired his experimental mindset. His understanding was warped, but he was brave in the face of its challenge. About a year after the conference, in December 2019, Hughes announced that he was going to make another launch up to the Kármán line, sixty-two miles (328,000 feet!) into the atmosphere. From there one would be able to see curvature, and I was excited to hear about the experiment. Just before his prior launch (in 2018), Hughes had said, "I expect to see a flat disk up there . . . I don't have an agenda. If it's a round Earth or a ball, I'm going to come down and say, 'Hey guys, I'm bad. It's a ball, OK?'"[21] Unfortunately, he never got the chance. On February 22, 2020, Hughes's rocket malfunctioned just after takeoff and he fell back to Earth and died. Say what you will about Hughes, but I will not criticize him. He embraced an adventurous spirit and core commitment to put his beliefs to the test, and promised to give them up if they did not pass, which is the foundation of the scientific attitude. But can the same be said of his Earth-bound fellow Flat Earthers?

In a delightful documentary called *Behind the Curve*, a film crew follows a group of Flat Earthers (most of whom seemed to be affiliated with FEIC) as they pontificate their views and occasionally try to test them. At first, the film might seem a celebration of Flat Eartherism, but once the characters are established, the fun begins. In one scene, a couple of Flat Earthers have spent $20,000 on a laser gyroscope to try to prove one of their core beliefs: that the Earth does not move. Except that when they turned on their equipment, they found a 15-degree-per-hour drift. Said one researcher, "Wow, that's kind of a problem. We obviously were not willing to accept that, and so we started looking for ways to disprove it was actually registering the motion of the Earth." They couldn't. Then—at the very conference I was attending in Denver—they were caught on film saying, "We don't want to blow this, you know? When you've got $20,000 in this freaking gyro. If we dumped what we found right now, it would be bad. It would be bad. What

I just told you was confidential."[22] Can one imagine an actual scientist saying this?[23]

As bad as this is, at the end of the film there is another experiment that is arguably worse. A group of Flat Earthers go out and try to measure whether a light beam lands at the same height on three equal poles that are spaced very far apart. Based on their theory, if the light beam hit the same height on each of the poles, this would prove that there was no curvature to the Earth. Actually, this is not a bad experiment, in that it is consonant with the famous Bedford Level experiment from the nineteenth century, which Alfred Russel Wallace (of evolution fame) set up to collect prize money to "prove" the curvature of the Earth.[24] So what did the Flat Earthers find? In the movie's final frame, we see them flummoxed because they can't get the light beam to go through the "right" hole on their apparatus. So they raise the pole. And the light goes through. Roll credits.

What was the result of all this experimental failure? FEIC 2019 went on as scheduled. As I said, for a Flat Earther there is no such thing as a definitive experiment. For all of their bluster about how much they care about evidence and paint themselves as more scientific than the scientists, the truth is that they don't really understand the basis of scientific reasoning. Their ignorance is not just about scientific facts, but about how scientists think. So how *do* Flat Earthers think? What is the basis (and weakness) of their reasoning strategy?

For one thing, their insistence on proof is based on a complete misunderstanding of how science works. With any empirical hypothesis, it is always possible that some future piece of evidence might come along to refute it. This is why scientific pronouncements customarily come with errors bars; there is always some uncertainty to scientific reasoning. This does not, however, mean that scientific theories are weak—or that until all of the data are in, any alternative hypothesis is just as good as a scientific one. In science, all of the data are *never* in! But this does not mean that a well-corroborated scientific theory or hypothesis is unworthy of belief. With science, it is ridiculous to ask for proof as a necessary standard.[25]

What scientists do often engage in, however, is *disproof*. If your hypothesis says that X must be true, and X is *not* true, then that means your hypothesis is wrong![26] For instance—as in the example from *Behind the Curve*—if a Flat Earther predicts no drift, and they actually found some, this means that their hypothesis is disproven. Now, of course, even scientists are allowed to go back and make sure that their equipment wasn't malfunctioning or that there isn't some other overlooked reason for the phenomenon they found upon experiment. But beyond a certain point, it seems ludicrous to keep making excuses. Given Flat Earthers' commitment to the power of proof, I am surprised by their cavalier attitude toward those experiments that have *disproven* their hypothesis.

Another weakness in Flat Earth reasoning has to do with their misunderstanding of how evidence gives warrant to a hypothesis. When a belief is warranted it means that it is justified on the basis of its evidence. The more evidence, the more credible the hypothesis. Of course, this falls short of proof. But does this mean that *no* amount of evidence can build up credibility for *any* belief, until the day comes when it is absolutely proven? If so, we would be justified in believing only the truths of math and deductive logic; both physics *and* Flat Earth would be thrown out the window. Yet to talk to a Flat Earther is to watch them shout "aha" with bright eyes any time they feel your failure to offer proof somehow makes their own hypothesis more credible. But this is just not how science works. To say that my hypothesis is unproven does not make yours more likely—else what about the triangular Earth, the trapezoidal Earth, or the donut-shaped Earth?[27] And of course the backtracking and revision, based on ad hoc rejection and groundless suspicions, just to keep their own hypothesis from being outright refuted, only undermines their credibility. This is not how scientists reason. One cannot keep modifying what one is willing to accept as evidence just to protect a favored hypothesis. Yet Flat Earthers routinely employ a double standard of evidence. Virtually anything a Flat Earther *wants* to believe is allowed to pass muster with hardly any scrutiny, whereas anything they do *not* want to believe is demanded to be proven?[28] But why?

I cannot emphasize enough how deeply Flat Earth is rooted in conspiracy theory reasoning. Indeed, some have described Flat Earth as the biggest conspiracy of all.[29] Time and again at FEIC 2018, I heard people talk about other conspiracy theories they believed: chemtrails, government control of the weather, fluoridated water as a means of mind control, the idea that the Sandy Hook and Parkland shootings were a hoax, that 9/11 was an inside job, and the list goes on.[30] One speaker actually said, "Everyone here can probably give you their top-twenty list of conspiracy theories." And indeed some confessed that because they were prone to conspiracy theories, *this is probably what drew them to research Flat Earth in the first place.* But the amazing thing is that they did not seem at all ashamed of this. One fellow explained this by saying, "Flat Earthers are more 'sensitive' to conspiracy theories than other people." But to believe that *all* world leaders are in on a secret that the world is flat? Does anyone think that Donald Trump and Boris Johnson could keep a secret like that? Apparently so. Time and again, Flat Earthers would come right out and tell me that belief in conspiracy theories was at the foundation of their reasoning.[31] (Indeed, in one of the seminars on how to recruit new believers into Flat Earth, one of the speakers said, "If you run into someone who says they don't believe in conspiracy theories, walk away.")

The specific role that conspiracy theories play in denialist reasoning will be covered in detail in chapter 2. For now, let me simply say that conspiracy-based reasoning is—or should be—anathema to scientific practice. Why? Because it allows you to accept both confirmation *and failure* as warrant for your theory. If your theory is borne out by the evidence, then fine. But if it is not, then it must be due to some malicious person who is hiding the truth. And the fact that there is *no evidence* that this is happening is simply testament to how good the conspirators are, which also confirms your hypothesis.

An equally large role in Flat Earth thinking is played by confirmation bias. Flat Earth is the ultimate example of motivated reasoning. They will cherry-pick or misinterpret any piece of evidence that will support their beliefs and reject with extreme bias any evidence that does not. As one of the five tropes common to all science denial reasoning, the problem of cherry-picking evidence too will be dealt with in chapter 2. Let me here simply point out that the mindset of virtually every Flat Earther I met at FEIC 2018 was to actively pursue anything that might tend to make their views seem more credible and ignore or dismiss anything that did not. Remember the reaction to the falsifying experiments in *Behind the Curve*? The idea of setting up a definitive experiment, and then living by the result, was anathema to them. They weren't even close to being scientists. They were true believers—evangelists for Flat Earth.

Naturally, I already had my suspicions about *how* Flat Earthers (and all science deniers) reason, but I still did not know *why*. If I hoped to be able to break through with Flat Earthers, and make them see that it was not just their facts but their reasoning strategy that was wrong, I needed to think a bit more about what might have led them to have this particular set of beliefs. Again, I felt out of my depth. Just as I am not a physicist, I am not a psychologist either. Yet based on my conversations so far, I did see a pattern in their stories that could perhaps shed some light on their motivation and mindset.

In addition to buttonholing speakers and some of the other superstars of Flat Earth, I also had a number of conversations with my fellow conference-goers. I found that if I got to an event early, when there were still lots of empty chairs, it was easy to strike up a conversation. One of the most interesting I had was with an older woman from Europe, who said that she was a documentary filmmaker. At first I was disappointed, as I suspected that perhaps she was not a believer in Flat Earth and just one of the folks like me who was here to observe the event. So I let my guard down.

"So you don't really believe all this stuff, then?"

"I don't believe it, I *know* it," she said.

Uh-oh, I had misjudged. Then, in the most pleasant way possible, she began to tell me her life story. She said that she used to be a scientist and had studied physics, chemistry, and psychology. But then she'd had a crisis in her life (which she did not specify, but I got the impression it was health related), after which her husband divorced her. She said this put her into a tailspin where she'd begun to question everything. What did her life mean? Who could she trust anymore? At this point she began to watch some Flat Earth videos and tried to debunk them but instead they had convinced her! She was embarrassed that she'd never questioned her "globalism" before, but said she'd had quite a regimented education.

At this point I said, "So could anything change your mind back?" After all, she'd changed her mind once, so I was curious what it might take to change it again. She said that nothing could. I probed a bit as to why this was the case and got a whiff that it might somehow be related to her religious beliefs. So I finally worked up the nerve to ask her another question.

"So, are you one of those folks who believe that God created the Flat Earth?"

"No," she said. "I don't believe that."

I thought that perhaps I'd run into my first nonreligious Flat Earther.

"So your belief in Flat Earth is secular?"

"No," she said. "I don't believe that either. Because I am the creator."

If she hadn't been so soft-spoken and pleasant, I might have thought she was joking. But it took me only a few seconds to realize that she was dead serious. She smiled and continued. She said that if God was separate from her, then she would be a victim. But that couldn't be, because she wasn't a victim anymore. So she must be God. She said that she had created the universe, and along with it the Flat Earth. She didn't buy into all of the other Flat Earthers who were talking about Christianity and Jesus. It was her!

With that she turned to an account of her present life and said that she'd moved back in with her husband—in America now—and that she was making films. She asked about me, and I told her I was a skeptic. That I didn't believe in Flat Earth. She said she was OK with this. I said that I had come to the convention to see what other people believed, and she liked this very much. She said to be careful, though. That she had studied indoctrination and felt that all globalists had been brainwashed! Rather than being mad or feeling insulted by my questions, she instead looked like she felt sorry for me. All during the presentation that followed—when we were sitting in nearby seats—she kept looking over at me and smiling when the speaker made a good point.

It was hard to keep my mind focused as I tried to process what I'd just heard. It would have been easy to dismiss this woman as crazy, but the weird part was that several of the things she'd said had resonated with things I'd heard from others. I am not saying that all Flat Earthers are delusional, but there was a common thread here that demanded follow-up. This woman had spoken of trauma in her life. And I now realized that several of the others I'd heard that day had spoken of a traumatic experience in their own lives as well, which coincided with the time they had started to believe in Flat Earth. For many it was 9/11. For others it was a personal tragedy. Some terrible event had occurred, which had caused them to do precisely what this woman had done: question everything. The conclusion she had come to—that she was God—had to be an outlier. But the idea that Flat Earthers were somehow drawn to the ultimate conspiracy theory at just about the time they were trying to heal from some important psychic wound was one that I just could not stop thinking about.

I had already noted that a number of Flat Earthers seemed alienated or marginalized from society. But that was easy to attribute to their belief in Flat Earth itself. As I said, Flat Earthers are often persecuted for their views and pay a heavy price with family, friends, community, and work. But now it occurred to me: What if they were alienated and marginalized to begin with? Maybe that is what led them to Flat Earth. Again, I am no psychologist, but something fell into place. If you were someone who felt that you were always on the outs in life, never quite fit in or had a chance, maybe never had the career or personal life you wanted, and felt that at least in part this was *because other people had been against you and lying to you and undermining you right from the start*, might it not seem attractive to explain all this through some giant conspiracy? Instead of being marginalized, suddenly you were part of an elite. You were one of the saviors of humanity, who actually knew a truth that billions of people had missed. And the fact that your cohort was so small only indicated the depth of the conspiracy against you. *The Matrix* indeed.

As I sat there, I concluded that perhaps Flat Earth wasn't so much a belief that someone would accept or reject on the basis of experimental evidence, but instead an *identity*.[32] It could give purpose to your life. It created instant community, bound together by common persecution. And perhaps it could explain some of the trauma and other difficulties you might have experienced in life, as the elites in power were all corrupt and plotting against you.

I will leave it to others to do the careful scientific work to measure the worth of these speculations.[33] But as a working hypothesis for myself, in that room at that time, it changed how I approached the rest of the conference. If I was correct, then Flat Earth wasn't really about *evidence* at all. The "evidence" was just a huge rationalization for one's social identity. This explained why Flat Earthers took it so personally when I challenged their beliefs. This wasn't just a belief that they happened to have; *this was who they were*. But this meant that I couldn't get them to change their beliefs without asking them to change their identity. And that sounded like a recipe for failure. How could I get anyone to begin to understand that their belief system was wrong without making it seem like I was attacking them as a person?

Perhaps I could proceed by taking them seriously *as human beings*, even while refusing to play their game of "proof." I could stop trying to offer my own evidence for the global Earth, but also refrain from asking for (or criticizing) their own. Instead I could engage them in conversation . . . about themselves. In this way, I thought that perhaps I could get Flat Earthers to do my work for me. For one thing, it would be disarming. But I also knew that my approach had to involve their reasons for believing in Flat Earth. Their beliefs were my entrée, but my goal was to get them to talk about why they had them.

Maybe I could ask them a question they'd never heard before. One that a scientist wouldn't have any trouble answering. And then—rather than trying to change their mind directly—I could just sit back and watch while cognitive dissonance overtook them, as they grew increasingly uncomfortable when they couldn't give me an answer.[34]

In his 1959 book *The Logic of Scientific Discovery*, Karl Popper offers his theory of "falsification," which says that a scientist always sets out to try to falsify their theory, not confirm it.[35] In my book *The Scientific Attitude*, I developed a key insight from this, which is that—in order to be a scientist—you have to be willing to change your mind on the basis of new evidence. So how about this for a question: "What evidence, if it existed, would it take to convince you that you were wrong?"

I liked this question because it was both philosophically respectable and also personal. It was not just about their beliefs but about them. So far, I had approached everyone at the conference with respect, and I planned to continue to do so. But now I would need to make a slight adjustment to my strategy. Instead of challenging them on the basis of their evidence, I would instead talk about the way that they were *forming* their beliefs on the basis of this evidence.

The next session was on "Flat Earth Activism" (in which they talked about how to recruit new members through street clinics to "wake people up") and was run by one of the biggest celebrities in Flat Earth. He was young and lean and had a look about him that was both intense and vulnerable. He was soft-spoken and patient, and obviously quite intelligent. Not only did he seem to be a true believer, but I gathered that a number of people at the conference believed in *him*. He was a natural leader, and that was a good thing, for he had one of the hardest jobs in Flat Earth, which was to convince people (sometimes face-to-face) to give up globalism.[36]

Immediately I was riveted. In a curious way, this man was setting out to do *precisely what I was trying to do*. I had come to the session to learn more about how Flat Earthers set about proselytizing new members. Perhaps I could learn some practical skills. He started off by showing a video of one of his street clinics, to demonstrate some of the techniques he used in trying to recruit people. His main piece of advice was that activists had to remain calm. Control their emotions. He said it helped not to assume that the people who believed in globalism were idiots or mentally ill. Give them your respect. Be upfront with them about your belief in Flat Earth, but also recognize that some people "aren't ready yet." There are so many lost people out there, he said. Don't expect to win every time. "You will face people who are in total denial of reality." (Yes, he actually said that.)

I had to smile. The tactics he was describing for recruiting someone into Flat Earth were not a bad script for how I hoped to bring them back out. If you just substituted "Flat Earther" for "globalist," he was describing almost every anecdotal account I had read of how people changed their minds and came to give up their resistance to vaccines or climate change.

From here, the speaker went on to share some standard fare about Flat Earth: water seeks its own level, the people at NASA have to sign nondisclosure agreements, all of the faked photos from NASA are taken underwater. Ho-hum. But then I saw a flash of anger as he began to describe the "purple pillars," which were people who believed in most other conspiracy theories but called Flat Earth people crazy. Were they heretics? That's what I think upset him. He was one of the folks who recognized—and did not apologize for—the role that conspiracy theories played in Flat Earth reasoning, and apparently felt that if someone was willing to believe that 9/11 was an inside job, or that the Parkland shooting was a hoax, they ought to be willing to come the whole way to Flat Earth. But he then recommended, for the sake of your own mental health, not to bring yourself to the point where everything in life is a conspiracy against you. He went on to make some personal remarks about his life and some ongoing medical issues—which I am not going to share here.

After the presentation was over, I felt reborn. This was the reason I had come to Flat Earth. Later that evening there was a scheduled Flat Earth "debate" between Rob Skiba and an alleged skeptic. Forget that. I wanted to have my own Flat Earth debate right now! I needed to talk to this fellow.

I waited patiently out in the hallway while things finished up, but when the speaker appeared—alone—I called him over and asked if I could take him out to dinner (my treat), on the condition that we spend the whole time debating Flat Earth. How could he refuse? Actually, many people would have, but I'd just witnessed quite an impressive performance and I had my hopes up that if I approached him in just the right way, he'd accept. I was honest and told him I was a skeptic. That I was a philosopher and scholar of science denial, and even that I was writing a book about this, but I would love to talk to him. To my delight, he accepted on one condition: that while I was trying to recruit him, he would try to recruit me!

We didn't have far to walk, since we decided to eat at the hotel restaurant. Just the two of us sitting across from one another at a small table. I asked him if I could take notes, and he said yes, and even offered that we could record the session if I cared to. I declined, feeling that this might interfere with our conversation. I didn't want either of us to have to "perform" for anyone but just have an honest, face-to-face encounter. He thought that was OK. We ordered our food, and then jumped right in.

I started by asking him to say a bit more about his life. It had been hard. He had a life-threatening medical condition and lived in a trailer, but was evicted from the property it had been on, and moved it to his mother's driveway. Her landlord had also made him move, and he ended up having to sell the trailer, which was painful because the Flat Earth community had taken up a collection to buy it for him. It wasn't clear where he lived now, and I didn't ask.

Now it was his turn. He seemed intrigued that someone like me would choose to come to a Flat Earth conference. He was wary (of course) but also disarmingly open and straightforward, and said that he wanted to ask me a question. "As an outsider who has now learned a bit about us: do you think that Flat Earth is ahead of its time?" I was worried that if I gave a straight answer, it would immediately put us at odds, so I said, "Let's come back to that at the end . . . I'm here to learn from you." We never did get back to the question, which is probably good because my answer would have been, "No, you're about five hundred years too late."

Then we got down to business. I knew I'd probably never get another chance like this. Here I was talking to a Flat Earther who was intelligent, genuine, and a very skilled debater. I even liked him. I didn't want to squander any goodwill and trust we'd built so far, but neither did I want to gamble that it would still be there later in our conversation, so I decided to start off with my most important question: "I understand that your view is compatible with belief in a creator, yet it doesn't seem to be faith-based. You guys are looking for evidence, which means that evidence must matter for your beliefs. So what specific evidence would it take to prove to you that your belief in Flat Earth was wrong?"

He gave me a pained expression. I don't think he'd ever heard that question before. But while his face furrowed, I saw his mind engage as he considered the question carefully. "Well, first, I'd have to be part of any experiment. Otherwise I wouldn't trust it." I said OK. He went on to speculate that perhaps a fully funded rocket that would allow him to go up sixty-two miles (to the imaginary point where space begins) would allow him to see for himself. I explained that bomber planes had gone up as high as 80,000 feet and could see the curvature of the Earth from there, but he said perhaps the window was curved, so how could he be sure?

We both sat for a minute with the idea of what it would mean to go to the edge of space and look out the window. He said that people in the Flat Earth movement loved him and that if he came back from a rocket trip and said that he no longer believed in it, they would be devastated. A lot of people would lose their belief. And, of course, it was also unrealistic to think that he would ever be able to go.[37]

I then proposed the earlier experiment I'd heard about in Skiba's seminar, where we might take a boat ride *way* out in Lake Michigan, beyond where it would be possible to see the superior mirage effect, and then look back at the shoreline.[38] Perhaps a hundred miles. If we still saw the Chicago skyline, then Flat Earth was right; if not, then it was wrong. It would be a definitive experiment. He didn't agree with this. He said there were too many variables based on weather and water vapor in the air. I said we could wait for whatever he might define as "perfect conditions," but he said no . . . too many variables.

I could see the struggle on his face. Just as much as I wanted to debunk Flat Earth, he wanted to be able to tell me what would count as definitive proof for him. He was smart enough to see the box my question had put him in: if he refused all evidence, it meant that maybe his beliefs were based on faith after all.

For a while he said nothing. Then I proposed that together we take a flight over Antarctica. I had heard several of the other speakers that day say that Antarctica was not a continent and that evidence of the conspiracy to cover this up was shown by the fact that there were no direct flights over Antarctica. At this point he said, "But there aren't any flights over Antarctica." I said, "Oh, no?" and reached into my back pocket, where I had come prepared with the itinerary for a direct flight from Santiago, Chile, to Auckland, New Zealand. If Flat Earth was right, such a flight shouldn't exist.[39]

"Did you ever take that flight?" he said.

"No, but here it is."[40]

He said he'd have to take the flight himself to believe it. If he could bring his own equipment and do any experiments he liked while onboard, then he would believe in a global Earth.

Wow! I was impressed. For the first time at this conference, I had gotten an answer to my hardest question. In a way, Mike Hughes had answered it by saying that if he went up to the Kármán line and saw a round Earth, he would give up his views. But the chance of that actually happening, on a homemade rocket no less, seemed wildly impractical. But here I was sitting across from a Flat Earther who was willing to come with me on a commercial flight that actually existed, and we could take it together.

The flight cost $800 per person. He said he didn't have the money. But how hard would it be for me to go home and set up a Facebook or GoFundMe fundraiser for all of my philosophical and scientific buddies to fund a trip like this? Wouldn't you chip in fifty bucks to watch a Flat Earther take a flight that he said didn't exist, then have to reckon with the consequences when that flight flew over Antarctica? I told him that I could fund this myself, probably by the time I got back to Boston.

Now my dinner companion was starting to look quite uncomfortable— and to tell you the truth, I was a bit worried too. Things were getting serious. If we were actually going to do this, I'd need to have some sort of reassurance that when it was all over he wouldn't say, "Well, the windows were curved," or something like that. And what were these experiments he wanted to perform? I didn't want to raise and spend $1600 of other people's money, only to have him back out at the end. We needed a criterion.

I gently offered that if we were serious about this, it was probably best to agree beforehand on what would count as a "successful" confirmation or refutation of Flat Earth. I proposed that a good measure might be whether we had to stop for fuel. If I was right, and Antarctica was a continent that was only about a thousand miles across, then we should be able to make the trip without stopping to refuel. In fact, if you think about it, the minute you stepped on the plane, it would be an enormous leap of faith, for if you did *not* believe we could make it, where could you stop in Antarctica for fuel? If, on the other hand, he was right that Antarctica was a mountain range—about 24,000 miles long—then we would never be able to make it on one tank. Even the longest nonstop flights can go only about 10,000 miles without stopping to refuel.[41] There *are* no around-the-world flights (even flying east to west) that can make it without stopping. So was that OK?

To my delight and amazement, he agreed. And we shook hands! I was brimming with excitement, for I knew quite clearly at this point that I had him. Maybe at some level he realized this too, for he slowly started to shake his head. "No, I can't," he said. "I take it back." "Why?" I asked. He said that perhaps fuel stops were an illusion. That maybe we were conditioned to think that planes needed to stop to refuel on all of these other flights, so that when the day came that we wanted to take one over Antarctica, we'd say that the Flat Earther had to stop for fuel. But what if we didn't? What if you could make it completely around the world on a single tank of jet fuel, and all of those other flights were just a shill to throw us off the scent?

I couldn't believe it.

"So let me get this straight," I said. "You're saying that you believe the entire history of jet travel, both in this country and around the world, has been a hoax since before you and I were born, to guard against the day when we would be sitting here tonight trying to come up with some criterion to measure whether the Earth was flat?"

He said yes.

At that point—for all intents and purposes—our dinner was effectively over.[42] His position had been demolished, and we hadn't even finished our entrées. Rather than getting up and leaving, however, I instead took a page from his seminar and kept my cool. If I'd left, it would have been rude. Plus I'd forfeit any chance to advance the dialogue. You don't change someone's mind by going back to your room and "being right." But I also felt the weight of Thomas Henry Huxley's admonition that "life is too short to occupy oneself with slaying the slain more than once." What should I do?[43]

I could see that he was a bit upset, so I moved things onto familiar turf and just let him talk for a bit. He asked if I was spiritual; I said no. He then went on to explain the relationship between God and the Devil and give me a mini-seminar on Flat Earth 101. At that point I was fine with that. I probed a bit, asking, "But if the devil is competent enough to hide such a big truth, why does he leave so many clues that you seem to have picked up on?" He explained that the truth is often hidden in plain sight. That people in control can control the narrative too. Just like what happened at the Parkland shooting.

My blood pressure jumped. My wife and I have a very good friend whose sister lost a child in the massacre at Sandy Hook. If I got angry, this dinner really would be over. But how could I let him get away with this hogwash? He began to talk about how the Parkland kids were "crisis actors." That the mom of one of the "victims" said, "I don't want thoughts and prayers, I want gun control," which made him a little suspicious. He said, "Isn't that exactly what the anti-gun lobby wanted her to say?" At that point our conversation devolved into a long back-and-forth over conspiracy theories and burden of proof, Occam's Razor, and why I had such a big problem with the idea that you could count speculations and suspicions as evidence. I made a calculated decision not to share that I knew a family who had been traumatized by this kind of nonsense. Later I regretted that. Maybe I should have blasted him. He wasn't the only victim in the world. Maybe he needed to hear that the kind of logic he was using had real consequences for real people.

By the time our plates were cleared—now past the second hour—we had returned to the topic of science denial. He said he didn't like how climate-change deniers and anti-vaxxers looked down on Flat Earthers. He was also upset about the "moral superiority" of scientists and made a case that if they were really scientists they should *want* to investigate Flat Earth. I told him that in science you had to earn your place at the table; that scientists didn't just go around looking into every conspiracy. "Well, I don't distrust science," he said. "I distrust pseudoscience." "Me too," I said. So we ended on a point of agreement.

As we got up to leave, I paid the check and he put a little Flat Earth brochure on the table for the waitress. We shook hands and parted on the best terms we could. He was a skilled, relentless debater who never gave an inch. It was jarring to me that he had so many unsubstantiated beliefs, and I wondered how any intelligent person could do that. People sometimes dismiss Flat Earthers as crazy or stupid, but I don't think that explains what's going on. Yes, there is deep ignorance of basic physics, and a heaping dose of willful ignorance and resistance at a level that may seem pathological, but the mindset was about something else. Here was a guy who knew enough rhetoric that he could counter (to his own satisfaction at least) anything I said. Of course he was wrong. But did he know that? And if he did, would he ever admit it? Probably not, but that didn't necessarily make him crazy either. For there were too many others out there just like him.

The argument he'd made over dinner had the same form as virtually every other denialist belief. Even if climate-change deniers and anti-vaxxers seem less extreme than Flat Earthers, they are using the same playbook. Even its own adherents would admit that Flat Earth is extreme. Some even wear that as a badge of honor. But I left thinking that it wasn't the specific content of Flat Earth beliefs that made it ridiculous. It was how they reasoned. And that wasn't unique to Flat Earth itself.

The actual debate on the main stage at the conference that evening was a bust. They had brought in a shill, and I left after ten minutes. He started by saying that he was a Catholic and that he'd been interpreting the Bible for forty-five years and accepted it as an authority "as far as it could go." Perhaps the Bible had no business making pronouncements about physics, he said, but he wouldn't know. I left at the point where he said, "Each of us have to be humbled before the word of God."

It was the end of the first day.

The following morning I had a brief conversation with Robbie Davidson, the conference organizer, when I passed him in the hallway. He didn't know that I wasn't a Flat Earther, so I asked, "I've heard a lot of researchers here say they don't have enough money for their experiments. You must be making a lot of money on this conference. Do you make any donations to them?" He replied, "I don't make a lot of money on these. These cost a lot to put up. My wife and I lost money on the first one." I pointed out that he had future conferences coming up, so perhaps he could do a fundraiser to get donations for some of the researchers. He said he'd take that into consideration.

Given the day I'd had yesterday, most of the sessions seemed like review. One after the other they all covered the same material. The only one that I was really looking forward to was called "Talking to Your Family and Friends about Flat Earth." Once again, I got there early. The session was run by two Flat Earth "researchers," both of whom seemed quite smug but promised to have different points of view: one was drawn to Flat Earth by Christianity and the other said he was secular.[44] The secular one said that he lived near the World Trade Center on 9/11 and had a view of it out his window. What he saw happening in real life wasn't what was being reported on the news. It made him start to doubt things. His belief in Flat Earth apparently followed soon after this, as he started to watch some online videos and tried to debunk them, but couldn't (and apparently assumed that if a person as intelligent as himself couldn't do it, they must be true). He said that his perspective was not Bible-based. It was based on "evidence." (I noted the familiar logic: the only criterion is proof. So if you can't prove that the Earth is round then it must be flat. QED.)

The next speaker said that his views *were* Bible-based and that he was drawn to Flat Earth because it aligned so well with his beliefs about the Bible. He'd never questioned 9/11 before he started to question the shape of the Earth. As with so many others at the conference, I surmised, Flat Earth was a gateway to other conspiracies. He went on to say that questioning global Earth led him to question NASA. "We're not taught how to think, we're taught what to think." He felt that we had been brainwashed, and that fluoride in the water only made it tougher to learn how to think. Here both speakers spoke favorably of the "red pill" scene in *The Matrix*, which brought a murmur of approval from the crowd. Everyone seemed to love that film. They were the people who knew the truth, and they were there to "wake the others," which is what today's session was all about.

They began with an interesting philosophical point: that there was a difference between causation and correlation. Having evidence that appeared to support something didn't amount to proof! The evidence in favor of global Earth doesn't prove it. It just correlates with it. But it also correlates (they said) with Flat Earth. So your job in talking to people about this is to get them to take that first step and begin questioning things. In fact, one of the most effective tactics is to let the person you're trying to convince ask *you* a question.

After some ridiculous "evidence" for Flat Earth having to do with an alleged collusion between Walt Disney and Wernher von Braun (who designed a rocket that was a precursor to the Apollo space mission), I was treated to the insight that if you looked closely at Walt Disney's signature, you could see that the loops hid three sixes in it! Of course, this didn't "prove" anything, I imagine, but there it was: evidence. So it had to be explained. And so on . . .

When the presentation finally returned to how to convince others to believe in Flat Earth, they said that not everyone could be converted. The speaker at yesterday's "debate," for instance, was a lost cause. "He's got too much to lose," one of the speakers said. "We'll never get him." Similarly, they offered that teachers and scientists were the hardest to convince because they were the most indoctrinated! A bit of practical advice was to walk away from anyone who said that they didn't believe in conspiracy theories. It wasn't worth your time. What was essential, though, was to know the details of the globalist system. Know how fast the Earth is (allegedly) spinning or rotating. They said that most globalists didn't even know their own system (which is probably true), so it was helpful to get them into areas where you "knew the facts." This was especially helpful in talking to strangers, because you were probably never going to see them again. But talking to friends and family could be the hardest thing of all.

The goal for a Flat Earth activist is to "plant the seed" of doubt, they said. Don't try to bulldoze people, especially family and friends. With strangers, make them commit to a length of time to discuss it. No hit and runs. Establish some ground rules, like "you can ask me a question, but then you've got to wait for me to finish." Burden of proof was essentially not an issue here. Their strategy was to get someone to question their own beliefs—or admit that they didn't know something—then see where that led.

The "secular" Flat Earther said, "If someone believes 9/11 as reported in the news, you've got a hard job ahead of you." What might work, though, is to recognize that even if you don't convince someone on the spot, you can plant a seed of doubt that comes to fruition later. Maybe ask people to research Flat Earth in private for two weeks, without telling anyone else what they're doing. After that, if they're convinced, they can share it with others.[45] Then I heard one of the most stunning pieces of advice I'd heard in the whole conference: that it was easier to have a relationship with someone you'd met through the Flat Earth community. "Look around at everyone in this community!" That got a big round of applause from the ballroom audience. It was as if they were trying to isolate themselves from others who might cause them to question their own beliefs.

Now it was time for the Q and A.

The first question was how to advocate for Flat Earth through your church. One fellow was getting hostility from his preacher. He was afraid of getting kicked out. The speakers' advice was to pursue others in the congregation. Perhaps put Flat Earth literature in the Bibles in the pews.

Second question: "What should I do if I am a Christian above all else, and I wonder whether my focus on Flat Earth is conflicting with my idea of teaching the Gospel? We are heading toward the end times. I need to be out there saving people." Answer: Try to infiltrate your congregation.

Third question: "What should I do if I'm out there as a Flat Earth activist and I'm in a group that is hostile to what I'm saying?" Answer: Lay down the rules. They can ask you one question at a time and then you get to answer. You don't want to be peppered with questions and then have someone get frustrated and say, "What does it matter?" and walk away.

At this point, one of the speakers shared that one of the most frustrating conversations he'd ever had was with a very polite man who just kept saying, "Yes, the Earth is flat, but why is it a perfect circle?" He'd explain, and then the guy would say again, "Yes, but why is it a perfect circle?" At this point I almost burst out laughing. If I ever encounter a Flat Earther in Harvard Square, now I know exactly what to say to them. The speaker shook his head and said, "Some people just don't want to learn."

Now it was time for the question that nearly broke me. Seriously. So far, for the most part, I'd been able to keep my cool—even at dinner the night before—but now I started to wonder if I was going to lose it. The question was asked by a man standing next to a little girl, about five or six years old. He said: "What can I do to keep my kid from getting bullied in school? We're grown-ups and we can take it, but she's being persecuted for her parents' beliefs." At this point my heart broke. Although I'd seen a couple of kids at the conference, I now felt the weight of the problem. Virtually all of the adults—by their own admission—used to be globalists and were converted through YouTube videos. And if they'd been converted once, perhaps they could be converted back. But what chance did you have if you were raised in a cult? If you grew up in a family where all you heard day after day was conspiracy and not to trust science?[46] That little girl never stood a chance.

My hands started to shake as I waited for the answer.

First the audience applauded the little girl for *standing up for her beliefs*. Then the speaker got a wicked smile on his face. "Kids are the best ones to go after," he said. Since the teacher was admonishing the child for bringing up Flat Earth in class, he advised her to go out and talk to the kids on the playground, where the teacher wasn't listening. "Some kids are willing to learn."

I looked around the room. The odds were a hundred to one against me. What would happen if I raised my voice and yelled, "Bullshit!"?

Instead, I got up and left the conference.

That night I didn't eat dinner with any of the Flat Earthers, and vowed to get out of the hotel. It was the last night of the conference anyway, and I didn't want to hang around for the awards banquet. So I just left and ate at a local restaurant.

While I did, my thoughts came to me thick and fast.

For everyone who thinks that Flat Earthers are harmless—and that the best way to deal with them is to just ignore them or laugh—I wondered if they knew what was coming. Based on what I'd seen, the Flat Earthers weren't just wrong, they were dangerous. They were organized and they were committed. And they were adding new members every day. The very fact that they'd had *two* sessions on recruiting new members—not to mention the convention itself—meant that they were expanding. They were taking up collections to buy billboards. They were courting celebrities. They were running street clinics to "wake people up." As such, they were at least a menace to science and education. But they were also contributing to a culture of denial that has gripped this country over the last few years, enabling hundreds of thousands of people to refuse to vaccinate their kids, politicians to refuse to take action on climate change, and gun-toting protestors to parade during a pandemic.

Not only that, I think that Flat Earthers are dangerous in their own right. Right now, most people laugh at them. But I defy you to go to one of their conferences and keep on laughing. We used to laugh at anti-evolutionists too. How many years before Flat Earthers are running for a seat on your local school board, with an agenda to "teach the controversy" in the physics classroom? If you think that can't happen—that it couldn't possibly get that bad—consider this: eleven million people in Brazil believe in Flat Earth; that is 7 percent of their population.[47]

I took two things from my time at FEIC 2018. First, I was right that the underlying reasoning of Flat Earthers was the same as that of climate-change deniers, anti-evolutionists, anti-vaxxers, and others. It wasn't just the content of their beliefs, but the reasoning process that had gotten them there that was corrupt. Ironically, I also learned a bit *from the Flat Earthers themselves* about what might be best to push back against them. Remain calm. Be respectful. Engage them in conversation. Try to build some trust. Say what you will about their beliefs and reasoning, they had the conversion tactics just right. In order to change someone's beliefs, you have to change their identity.

As I got ready to fly home the next day, I had more time to reflect. Yes, I'd learned something about how to talk to a science denier, but had I made a real dent in even one of the Flat Earthers' beliefs? Well, how would I know? No, I didn't convert anyone. No one tore off their lanyard and followed me out to the parking lot. But was that the criterion? And was that the point? I went to FEIC 2018 not to change minds but to understand better how their minds worked. I would dearly love to have had more influence, but there are no magic words you can say to convert someone on the spot, especially in a crowd of their peers, when they have come to a conference expressly for the purpose of reinforcing their identity.

And hadn't I at least planted the seed of doubt a few times? When I buttonholed Skiba coming off the stage, we drew quite a little audience. When I had dinner with the other speaker, I'd given him plenty of reason to doubt, even if he didn't listen. Bringing someone back from a belief like this is probably a long game. It takes a while to build up trust. I couldn't just go once, tell them the truth, and expect a miracle. But at least I had shown up. That had to count for something. And what if, in future years, more people did what I tried to do, once they were aware of the problem?

Sitting in the departure lounge at the Denver airport, I spied a pilot for a major commercial airline walk past. Suddenly I was in *The Matrix*. Did he know? Was he in on the conspiracy? It was weird. I'd spent the last forty-eight hours surrounded by people who believed in an incredible conspiracy that the Earth was flat. And here I was surrounded by people who almost certainly did *not* believe that. But who could tell? Curiously, even though I was back in civilization again, I still felt isolated. I felt infected. Perhaps there was another Matrix . . .

I hustled a bit and caught up with the pilot, who was leaning against a pillar, texting on his cell phone. "Do you mind if I ask you a question?" I said. He nodded, but surely had no idea what was coming. "I just got back from two days at a Flat Earth convention. Now, don't worry, I'm not one of them. I'm a scholar who was there to study how they end up believing such a crazy thing. But a couple of the speakers said things about air travel and the curvature of the Earth that I know are wrong, so can I ask you a question?"

I'm not sure he completely believed me. Even if I was telling the truth, it was a lot to take in at once. But he nodded and said, "Sure." We both had a little while before our flights.

He said they were right that the compass does funny things over the South Pole. There was some literature on this that he said he'd send me (which he did). But they were wrong about flights over Antarctica. There was one that he knew of, but there weren't a lot. The problem was that under aviation regulations you could only fly a route like that in a 777 or above, because you had to be within a few hours of a "ditch" site, where you could land in an emergency. This meant that even if the fastest route between South America and Australia was over Antarctica, for commercial travel they probably weren't going to take that route.[48]

When I asked about seeing the Earth's curvature in flight, he smiled. "Not at thirty thousand feet. I've heard that some of the bombers go up to sixty thousand feet. At that point you can see curvature. But I've never seen it myself."

"So you're not in on the conspiracy, then?"

"No," he said with a smile. "I guess not."

We traded business cards and later exchanged a few emails. I apologized for the weird questions, and hurried away to catch my flight. But I suspect I made his day. Now he had a story to tell.

By the time I landed in Boston, I felt much better. I was home. The last two days had felt like a month, but I was out from under now. It was worth going, but also strangely stressful. I'd had a few moments of unreality along the way where I'd ask, "Is it me or is it them?" I headed for the men's room before going to get my luggage. As I locked the stall behind me, I looked over at the wall and saw this piece of graffiti (I kid you not): "The Earth is Flat."

It would have been smart to end the chapter here, but that's not where the story ends. When I got back home, I found that I was a minor celebrity, with all of my stories and observations. People at parties would crowd around and make me tell it again about my time at Flat Earth. I already knew that I was going to write about this in a book, but there was so much immediate interest that I decided I couldn't wait. Seven months later I had the cover story of *Newsweek*—June 14, 2019—with the astonishing title "The Earth Is Round."[49]

After that, I did a few radio shows and some other publicity, which led to lunch with a local physicist who'd heard me on NPR. He invited me to do an opinion piece called "Calling All Physicists" for the *American Journal of Physics*.[50] In it I told my story (again) but also put out a plea for more scientists to take Flat Earth seriously. I'd spent two days talking about reasoning strategy, but asked for someone with some physics training to please come help me crash the next Flat Earth conference and do some "content rebuttal."

Amazingly, I got an offer. Bruce Sherwood is a retired physicist who lives in Texas. He and his wife, Ruth Chabay, are the authors of one of the leading textbooks on how to teach physics using computational modeling. Bruce was patient, focused, and completely fascinated with the stories I told him. Better than that, he took them seriously enough that at several points he said, "That's interesting," and promised to go off and do some research.

After several rounds of questions—involving me and one of his collaborators, Derek Roff—one day he announced that he had built a 3-D computer model of Flat Earth!

I couldn't believe it. As I examined the model, he explained that it would allow Flat Earthers to explore their own system, and see if what they were predicting was consistent with their own theory. Of course it wasn't. For instance, if they were right that Antarctica was a mountain range at the perimeter of the Earth, what did that mean for their view of different stars?

"Walk inside the model and look up," Bruce said. "If you're standing at the North Pole, then Polaris should be directly overhead. Fair enough. But if you're standing at the 'edge of the Earth'—and Polaris is only a few thousand miles overhead—shouldn't you at best see it at an angle? But if you're actually in Antarctica, you won't be able to see it at all. Their model is inconsistent with physical observation. And they can see that for themselves."

Here is a link to the model.[51] Try it for yourself.

What genius to design something that takes Flat Earth seriously *and* conforms to their demand for firsthand observational evidence. The model may not prove global Earth, but it *does* disprove Flat Earth, or at least the model that those at FEIC were advocating. How do they explain its inconsistencies? Now the burden of proof is back where it belongs.

And here is the best part. At the next available opportunity, Bruce and I are going to go to a future FEIC convention, rent a booth at the swag fair, and invite people to try out his model.[52] We'll both be there—a physicist and a philosopher side-by-side—to engage in both content rebuttal and technique rebuttal. As the Flat Earth activists said themselves, it's not just about one conversation. It's about staying calm and building trust. And for that you need to keep showing up.

Who knows if we'll actually end up convincing anyone. But wouldn't it be a kick if my dinner companion showed up again?

2   What Is Science Denial?

After spending enough time around Flat Earthers, anti-vaxxers, intelligent designers, and climate change deniers, one begins to sense a pattern. Their strategies are all the same.[1] Although the content of their belief systems differs, all science denial seems grounded in the same few mistakes in human reasoning. This has been studied by previous researchers such as Mark and Chris Hoofnagle, Pascal Diethelm and Martin McKee, John Cook, and Stephan Lewandowsky, who have come to consensus on five common factors:[2]

(1) Cherry-picking evidence

(2) Belief in conspiracy theories

(3) Reliance on fake experts (and the denigration of real experts)

(4) Committing logical errors

(5) Setting impossible expectations for what science can achieve

Together, these provide a common blueprint for science deniers to create a counter-narrative on any topic where they wish to challenge the scientific consensus. The Hoofnagle brothers define science denial as "the employment of rhetorical tactics to give the appearance of argument or legitimate debate, when in actuality there is none."[3] Why would anyone want to do this? Perhaps for self-interest. Or ideology. Or to conform to a set of political expectations. There are many reasons why someone might wish to create—or might be taken in by—a false reality, when the scientific consensus challenges what they would prefer to believe. We'll get to that. First, I would like to examine each of the five reasoning errors in more detail, so that we have a better understanding of *how* denialism is a problem for empirical judgment. Later, I will have more to say about *why* it might be that a common script lies behind it . . . and what we can do about it.

Of course, reliance on fake experts, illogical reasoning, and insisting that science must be perfect all seem fairly straightforward, don't they? It is easy to see what is wrong with them. But what about the problem of cherry-picking evidence? Or belief in conspiracy theories? These go to the heart of scientific judgment, which is supposed to be embedded in a good-faith effort to test one's theory against reality instead of just trying to confirm what one already wants to believe, or jumping to a conclusion based on no evidence whatsoever. Scientists set out to find the truth, not deny it when it doesn't conform to their expectations. If an ideologue is completely committed to a theory—dismissing any evidence against it and needing little in its favor—how will they learn from future experience?

It probably won't surprise you to learn that the flawed reasoning strategy employed by science deniers is firmly rooted in a misunderstanding of how science actually works. In my book *The Scientific Attitude*, I went over some of these misconceptions in detail. I won't repeat them here, except to say that one of the hallmarks of science is how it responds to evidence. Scientists *care about evidence* and are *willing to change their minds* based on new evidence. This is why science cannot offer proof, but must instead rely on the idea that belief is warranted when a theory has sufficient credible evidence and has survived rigorous testing.[4] With ideology or dogma, it's another story.

The Elements of Science Denial

As we saw in chapter 1, the five tropes of science denial reinforce one another. No science denier stops to use the tactics one by one, but instead moves seamlessly from conspiracy theory, to red herring, to questioning your experts or evidence in the manufacture of a seamless web of doubt. Still, it is worth going over each of the five tropes individually. This will not only establish that each of these errors can be found in the only example of science denial that we have explored so far—Flat Earth—but set the foundation for recognizing them in the other examples of science denialism that we will encounter later: climate change, GMOs, and coronavirus. As stated, our goal here is to show that all science denial uses a common pattern. Later, we will explore why.

Cherry-Picking

If someone wants to claim that their pseudoscientific theory has scientific merit, selectively choosing evidence in its favor might seem an attractive strategy. To claim that you believe a fringe hypothesis solely on the basis of *faith* just doesn't sound very scientific. To say that you have actual evidence sounds better. For the cherry-picker, though, it matters very much which evidence you choose: you must consider only that evidence that supports your hypothesis and ignore or dispute the rest, else your theory might be refuted.

We saw this tactic in use by the Flat Earthers when they noted that you can *sometimes* see the city of Chicago from forty-five miles out in Lake Michigan. What they neglect to mention is that on many days, you cannot see this. Surely the former demands an explanation. But so does the latter. When you try to explore this, however, the Flat Earther makes clear that they are *only* interested in the fact that Chicago is *sometimes* visible (which is consistent with their theory that the Earth is flat), and not at all curious about why it is sometimes *not* visible (which their theory fails to explain). In fact, as we saw, they will reject as fake any credible scientific theory that explains *both* why the city is sometimes visible and sometimes not, in favor of their own theory, which cannot account for why the city is not *always* visible.

This is a perfect example of the kind of selection bias that is at the heart of cherry-picking evidence, which is deeply rooted in a common cognitive error called confirmation bias.[5] With confirmation bias, we are motivated to find facts that are consistent with what we prefer to believe, and all too ready to ignore any facts that don't. Climate change deniers, for instance, sometimes insist that the global temperature did not go up during the seventeen years between 1998 and 2015, only because they cherry-picked 1998 as their base year (which had an artificially high temperature due to El Niño).[6]

The problem here is one of bad faith. Of not looking for evidence to *test* your proposal but only to *confirm* it. But this is quite simply not how science works. Scientists do not merely look for support for what they hope to be true; they design tests that can show whether their hypothesis might be *false*.[7] Although crucial experiments may be few and far between, it is the corrupt *attitude* of trying to confirm your hypothesis, rather than rigorously test it, that reveals the problem with cherry-picking evidence. With cherry picking, we are more likely to prop up a sketchy hypothesis that might long ago have been refuted, had we looked at the full case of evidence.

Still, this does not stop most science deniers, who insist that scientists are biased because they will not bring their work to a grinding halt and consider all of the cherry-picked facts that nonscientists have gathered. At FEIC 2018, I ran across numerous people who thought that they were entitled to kick in the door of science and say, "Look at these one hundred points that science cannot explain!" Yet even if I had the patience to try, and went down the list one by one knocking out the scientific explanation for ninety-nine of them, the typical Flat Earther would say, "Aha, but what about that last one!" Which is to say that they are unscrupulously selective. And they do not care about refutation.[8]

Conspiracy Theories

Belief in conspiracy theories is one of the most toxic forms of human reasoning.[9] This is not to say that real conspiracies do not exist. Watergate, the tobacco companies' collusion to obfuscate the link between cigarette smoking and cancer, and the George W. Bush–era NSA program to secretly spy on civilian Internet users are all examples of real-life conspiracies, which were discovered through evidence and exposed after exhaustive investigation.[10] By contrast, what makes conspiracy theory *reasoning* so odious is that *whether or not there is any evidence,* the theory is asserted as true, which puts it beyond all reach of being tested or refuted by scientists and other debunkers. The distinction, therefore, should be between *actual* conspiracies (for which there should be some evidence) and conspiracy *theories* (which customarily have no credible evidence).[11]

We might define a conspiracy theory as an "explanation that makes reference to hidden, malevolent forces seeking to advance some nefarious aim."[12] Crucially, we need to add that these tend to be "highly speculative [and] based on *no* evidence. They are pure conjecture, without any basis in reality."[13] When we talk about the danger of conspiracy theories for scientific reasoning, our focus should therefore be on their nonempirical nature, which means that they are not even capable of being tested in the first place. What is wrong with conspiracy theories is not normally that they have already been refuted (though many have), but that thousands of gullible people will continue to believe them *even when they have been debunked.*[14]

If you scratch a science denier, chances are you'll find a conspiracy theorist. Sadly, conspiracy theories seem to be quite common in the general population as well. In a recent study by Eric Oliver and Thomas Wood, they found that 50 percent of Americans believed in at least one conspiracy theory.[15] This included the 9/11 truther and Obama birther conspiracies, but also the idea that the Food and Drug Administration (FDA) is deliberately withholding a cure for cancer, and that the Federal Reserve intentionally orchestrated the 2008 recession. (Notably, the JFK assassination conspiracy was so widely held that it was excluded from the study.)[16] Other common conspiracy theories—which run the range of popularity and outlandishness—are that "chemtrails" left by planes are part of a secret government mind-control spraying program, that the school shootings at Sandy Hook and Parkland were "false flag" operations, that the government is covering up the truth about UFOs, and of course the more "science-related" ones that the Earth is flat, that global warming is a hoax, that some corporations are intentionally creating toxic GMOs, and that COVID-19 is caused by 5G cell phone towers.[17]

In its most basic form, a conspiracy theory is a nonevidentially justified belief that some tremendously unlikely thing is nonetheless true, but we just don't realize it because there is a coordinated campaign run by powerful people to cover it up. Some have contended that conspiracy theories are especially prevalent in times of great societal upheaval. And, of course, this explains why conspiracy theories are not unique to modern times.

As far back as the great fire of Rome in 64 AD, we saw conspiracy theories at work, when the citizens of Rome became suspicious over a weeklong blaze that consumed almost the entire city—while the emperor Nero was conveniently out of town. Rumors began to spread that Nero had started it in order to rebuild the city in his own design. While there was no evidence that this was true (nor for the legend that Nero sang while the city burned), Nero was apparently so upset by the accusation that he started his *own* conspiracy theory that it was in fact the Christians who were responsible, which led to the prevalence of burning them alive.[18]

Here one understands immediately why conspiracy theories are anathema to scientific reasoning. In science, we test our beliefs against reality by looking for disconfirming evidence. If we find only evidence that fits our theory, then it *might* be true. But if we find any evidence that disconfirms our theory, it must be ruled out. With conspiracy theories, however, they don't change their views *even in the face of disconfirming evidence* (nor do they seem to require much evidence, beyond gut instinct, that their views are true in the first place). Instead, conspiracy theorists tend to use the conspiracy *itself* as a way to explain any lack of evidence (because the clever conspirators must be hiding it) or the presence of evidence that disconfirms it (because the shills must be faking it). Thus, lack of evidence in favor of a conspiracy theory is in part explained by the conspiracy itself, which means that its adherents can count both evidence *and* lack of evidence in their favor.

Virtually all conspiracy theorists are what I call "cafeteria skeptics." Although they profess to uphold the highest standards of reasoning, they do so inconsistently. Conspiracy theorists are famous for their double standard of evidence: they insist on an absurd standard of proof when it concerns something they do *not* want to believe, while accepting with scant to nonexistent evidence whatever they *do* want to believe. We have already seen the weakness of this type of selective reasoning with cherry-picking evidence. Add to this a predilection for the kind of paranoid suspicion that underlies most conspiracy-minded thinking, and we face an almost impenetrable wall of doubt. When a conspiracy theorist indulges their suspicions about the alleged dangers of vaccines, chemtrails, or fluoride—but then takes any contrary or debunking information as itself proof of a cover-up—they lock themselves in a hermetically sealed box of doubt that no amount of facts could ever get them out of. For all of their protests of skepticism, most conspiracy theorists are in fact quite gullible.

Belief in the flatness of the Earth is a great example. Time and again at FEIC 2018, I heard presenters say that any scientific evidence in favor of the curvature of the Earth had been faked. "There was no Moon landing; it happened on a Hollywood set." "All the airline pilots and astronauts are in on the hoax." "Those pictures from space are Photoshopped." Not only did disconfirming evidence of these claims *not* cause the Flat Earthers to give up their beliefs, it was used as more evidence for the conspiracy! And of course to claim that the devil is behind the whole cover-up about Flat Earth . . . could there *be* a bigger conspiracy theory? Indeed, most Flat Earthers would admit that themselves.

A similar chain of reasoning is often used in climate change denial. President Trump has long held that global warming is a "Chinese hoax" meant to undermine the competitiveness of American manufacturing.[19] Others have contended that climate scientists are fudging the data or that they are biased because they are profiting from the money and attention being paid to their work. Some would argue that the plot is even more nefarious—that

climate change is being used as a ruse to justify more government regulation or takeover of the world economy. Whatever evidence is presented
to debunk these claims is explained as part of a conspiracy: it was faked,
biased, or at least incomplete, and the real truth is being covered up. No
amount of evidence can ever convince a hard-core science denier because
they distrust the people who are gathering the evidence.[20]

So what is the explanation? Why do some people (like science deniers)
engage in conspiracy theory thinking while others do not?[21] Various psychological theories have been offered, involving factors such as inflated
self-confidence, narcissism, or low self-esteem.[22] A more popular consensus
seems to be that conspiracy theories are a coping mechanism that some
people use to deal with feelings of anxiety and loss of control in the face
of large, upsetting events. The human brain does not like random events,
because we cannot learn from and therefore cannot plan for them. When
we feel helpless (due to lack of understanding, the scale of an event, its personal impact on us, or our social position), we may feel drawn to explanations that identify an enemy we can confront. This is not a rational process,
and researchers who have studied conspiracy theories note that those who
tend to "go with their gut" are the most likely to indulge in conspiracy-
based thinking. This is why ignorance is highly correlated with belief in
conspiracy theories. When we are less able to understand something on the
basis of our analytical faculties, we may feel more threatened by it.[23]

There is also the fact that many are attracted to the idea of "hidden
knowledge," because it serves their ego to think that they are one of the
few people to understand something that others don't know.[24] In one of
the most fascinating studies of conspiracy-based thinking, Roland Imhoff
invented a fictitious conspiracy theory, then measured how many subjects
would believe it, depending on the epistemological context within which
it was presented. Imhoff's conspiracy was a doozy: he claimed that there
was a German manufacturer of smoke alarms that emitted high-pitched
sounds that made people feel nauseous and depressed. He alleged that the
manufacturer knew about the problem but refused to fix it. When subjects
thought that this was secret knowledge, they were much more likely to
believe it. When Imhoff presented it as common knowledge, people were
less likely to think that it was true.[25] One can't help here but think of the six
hundred cognoscenti in that ballroom in Denver. Out of six billion people
on the planet, they were the self-appointed elite of the elite: the few who

knew the "truth" about the flatness of the Earth and were now called upon to wake the others.

What is the harm from conspiracy theories? Some may seem benign, but note that the most likely factor in predicting belief in a conspiracy theory is belief in another one. And not all of those will be harmless. What about the anti-vaxxer who thinks that there is a government cover-up of the data on thimerosal, whose child gives another measles? Or the belief that anthropogenic (human-caused) climate change is just a hoax, so our leaders in government feel justified in delay? As the clock ticks on averting disaster, the human consequences of the latter may end up being incalculable.

Reliance on Fake Experts (and the Denigration of Real Experts)

One of the hallmarks of science denial is the idea that, until a theory is 100 percent "proven" (which no theory can ever be), everything is open to debate. This is meant to imply that, in the absence of complete consensus, it is justified to prefer the opinion of some experts over others. And guess which ones the science deniers are going to choose?

As we have seen, the point of science denial is to create a counternarrative to challenge the scientific consensus on matters that clash with their preferred ideology. Even if all (or most) scientists agree that cigarette smoking causes cancer, or that climate change is real, isn't it possible to sow a little doubt?[26] Whether this is achieved by cooking up crackpot theories (or theorists) or finding them in the wild, it doesn't really matter. The point is not to change the minds of actual scientists but instead to lobby for attention from the *audience* for scientific information, which often cannot tell one expert from another. The goal is to make it *look* like there is a debate, even when there isn't one. When science seems equivocal, or a result looks controversial, denialists win.

In his book *The Death of Expertise*, Tom Nichols outlines the problem of treating factual, empirical issues as if they were open to the sort of partisan lobbying and polarized squabbling that characterizes our political disagreements, which are "rooted in conflict, sometimes conducted as respectful disagreement but more often as a hockey game with no referees and a standing invitation for spectators to rush onto the ice."[27] This is exactly what science deniers seek to do with scientific matters: make them into ideological ones.[28]

This is most effectively done if it can be shown that the "experts" are biased. If the scientists who are claiming that climate change is real (for instance) are found to be liberals, or university trained, or grant funded, doesn't that raise questions about whether one might mistrust their motives and so also their conclusions? The populist mistrust of experts that Nichols discusses opens the door for science deniers and other ideologues to promote their *own* brand of experts who—even if you want to argue that *they* might also be biased—are at least holding up the other side in an open scientific controversy, which achieves a certain kind of balance that sounds fair-minded to uninformed outsiders who just want science to be "objective." But this, of course, leads to a sort of false equivalence, whereby science deniers feel justified in trusting their own "experts"—even if they have no expert training at all—against those whom they feel are biased against them.

As noted, this type of reasoning was on full display at FEIC 2018. When Robert Skiba took the stage and said that he had no scientific training but he *did* have a white lab coat, what could this be other than an attempt to show favor for some kinds of "experts" and denigrate others, whose only claim to authority is how they are dressed. What does this achieve? The stated reason is that the "other side" consists of experts who are all either biased or "shills" for a particular point of view; we can't trust them because they are paid off or otherwise corrupted from telling the truth. There is a profound sense of victimhood behind most science denial, whereby adherents complain that the so-called "real" scientists won't take them seriously or consider their own experts' data.

Conspiracy theories of course play a role here, as does the tactic of cherry-picking. The five tropes of science denial work in concert with one another in such a way that both reliance on fake experts and dismissal of true experts are not merely defining characteristics of science denial, but the inevitable outcome of belief in conspiracy theories, having impossible expectations for science, and all of the other tropes. It is a self-stoking cycle. The fake experts provide cherry-picked (or simply made-up) "evidence" that is used to question the consensus of scientific reasoning. When this evidence is not taken seriously, suspicions may grow and tribal reasoning takes over. Scientific disputes begin to resemble political ones, in a war of "us against them." Once the other team has been demonized sufficiently, it is easy to look for clues that—to a certain mind—might suggest a conspiracy, which further justifies reliance on one's own experts as opposed to any other.

The whole thing is built around trust—or rather the lack of it—which makes the dispassionate, objective evaluation of evidence to settle scientific disputes into an impossibility.

Or at least it throws the whole process into doubt. And doubt is all that a science denier needs.

Illogical Reasoning

There are myriad ways to be illogical. The main foibles and fallacies identified by the Hoofnagle brothers and others as most basic to science denial reasoning include the following: straw man, red herring, false analogy, false dichotomy, and jumping to a conclusion.[29]

I would be shocked if most science deniers had taken a course in informal logic. They likely have no training at all in understanding the above fallacies or even know their names. Yet they are experts in their practical use. When the climate change denier says that "carbon dioxide is not the *only* driver of climate change," that is a perfect example of the straw-man fallacy, where one imagines the weakest version of an opponent's argument because it is easiest to knock over. Virtually no responsible climate scientist would deny that there are many possible drivers of climate change, including natural ones. But that is not the point. Right now human-caused carbon dioxide emissions are by far the largest and fastest-growing cause of global warming. But the climate change denier doesn't want to talk about this.[30] So instead they invent a straw man, even though nobody said that human activity was the *only* cause of global warming.

Likewise, when Flat Earthers say, "Did you know that there are three sixes in Walt Disney's signature?," what else could this be but a red herring? Yes, the sixes are there for anyone to see. (In fact, once you see them it is maddening that you cannot *unsee* them.) But what is that supposed to prove? That Walt Disney was in on a conspiracy to cover up the truth about Flat Earth? If so, some actual evidence must be provided to connect the dots. The signature alone is irrelevant to the shape of the Earth.

By relying on such specious arguments, science deniers are engaging in a series of reasoning errors that have been identified, examined, and refuted by logicians and philosophers over the last 2,300 years. This is not the time or place to give a long (or even a short) course on logic.[31] It is also not the place to enumerate endless examples of the illogical reasoning of science deniers. If one desires more material to show that these and other logical fallacies are at the heart of Flat Earth, anti-vaxxer, anti-evolution, or climate change denier reasoning, there are a host of excellent resources.[32] In later chapters, I will also provide more examples.

Insistence That Science Must Be Perfect

The only people who insist that science must be perfect are those who have never done any science. Nonetheless, we often hear science deniers hold out for impossible standards by saying things like, "Can you prove that vaccines are 100 percent safe?" or "Why don't we wait for the rest of the evidence to come in on global warming?" or "The causal link between cigarette smoking and lung cancer has never been conclusively established." As previously discussed, this is not mere skepticism but the kind of ideologically motivated denial that someone brings up when they don't want to believe what the overwhelming consensus of empirical evidence is telling them.

Given the nature of inductive reasoning, there will always be some residual uncertainty at the base of any scientific hypothesis. Short of abandoning the foundational idea that new evidence can always come along to help us modify or even overthrow a scientific theory, we cannot expect science to conform to the same standard of proof and certainty found in mathematics or deductive logic. Yet in the hands of a science denier, any scintilla of doubt can be exaggerated to pretend that there is a debate even when there isn't one.

Science deniers routinely exploit the uncertainty of science. As shown, they are infamous for their double standard of evidence. No amount of evidence could possibly convince a science denier to believe something that they do not want to be true; they will insist on proof. Yet little evidence is needed to convince them that their own hypothesis is credible, for they trust their own sources. This is an outright perversion of the rational basis for science. There is no need to prove that something is certain for it to be believable. In science, there is such a thing as "warrant," which means that if there is enough evidence in a theory's favor—and it has been rigorously tested to see if it can be refuted—then there is rational basis for believing it to be true *even if* we must always hold out the possibility that some future evidence may later overthrow it.[33]

To deny this is effectively to claim that we *cannot know anything* about the empirical world until all of the evidence is in. Which means never. To a smug science denier, that might seem fine. But are they really willing to throw out *all* scientific beliefs, along with the ones they denigrate? Yes, we would suddenly have no basis for believing Darwin's theory of evolution by natural selection. But neither would we have warrant for their preferred alternative of intelligent design, nor for antibiotics, transplants, or gene editing. True, the basis for anthropogenic climate change would be undercut, but so would all weather forecasting, the chart of tides, and the science behind agriculture.

The problem with cafeteria skepticism is that it pushes things toward a ludicrous inconsistency. How did the Flat Earthers justify using cell phones to live tweet FEIC 2018, when part of their cellular traffic bounced off satellites?[34] What to say about the homeopathy enthusiast who makes a deathbed conversion and suddenly decides that he wants chemotherapy after all? These people actually trust science, just not the kind of science they prefer not to believe in. But how can this be anything other than ludicrous?

Another absurdity behind the science denier's insistence on ideal standards for science is inference that until the day comes when proof is offered for Darwin's theory of evolution by natural selection, or for global warming, any theory is just as good as any other. We've all heard the creationist say that evolution is "only a theory." But intelligent design is only a theory too. But why, then, they might want to know, don't we investigate both and perhaps "teach the controversy" in the biology classroom?

Here their misunderstanding is not merely about certainty; it's also about probability. Remember that the concept of warrant is rooted in the idea that one's belief in a scientific hypothesis is *proportionate to the evidence in its favor.* With something like Darwin's theory of evolution by natural selection, it is so well corroborated by 150 years of scientific experience that it forms the basis for virtually *all* of our understanding of biology. Evolution by natural selection is the backbone for genetics, microbiology, and molecular biology. In a 1973 essay, the eminent biologist Theodosius Dobzhansky said that "nothing in biology makes sense except in the light of evolution."[35]

But still, the science denier might claim, wouldn't science be improved if we held out for certainty? After all, the iconoclast is sometimes right. They laughed at Galileo, didn't they?[36]

But really? Do you want to play that game?

In February 2019, Reuters published a story that said that the evidence for human-caused global warming had reached the "gold standard" of confidence, at the "five-sigma" level. This means that there is only a one-in-a-million chance that the climate change deniers could be right. Granted, this is not the same as certainty. But it is the same level of confidence that was reached in 2012 to announce the discovery of the Higgs boson subatomic particle, which is a basic building block of the universe.[37] Of course, someone might continue to doubt the evidence, and claim that you should embrace the science denier's alternative because it "might" be right. But that is an absurd standard for belief.[38]

How does one shame the shameless? In the face of ridiculous belief, perhaps ridicule works best. Do you remember the classic scene at the end of the 1994 movie *Dumb and Dumber*, where Jim Carrey's character is desperately trying to convince a woman to go out with him? He tries everything and she keeps turning him down. Finally, he asks her to assess the probability that he'll be able to convince her to go on a date. "One in a million," she says. At which points he smiles and says, "So you're telling me there's a chance . . ."

You don't want to be this guy.

The Motivational and Psychological Roots of Science Denial

Now that we understand the tactics behind science denial, another important set of questions arise. How did all of this happen? What are its origins? And can any of this explain why science deniers all seem to be working from the same script? In short, the challenge is this: if the five traits outlined are such a bad form of reasoning, why are they so widespread?

Here it is important to draw a distinction between two possible methods of approach. One is to focus on the way that science denial is *created*. The other is to focus on why people *believe* it. The latter has customarily gotten most of the attention, and led to the popular—but overly simplistic— idea that science denial is due to simple ignorance. But that can't be the whole story, even for why people believe it. (Indeed, research has shown that some of the staunchest science deniers are the most educated.)[39] And it certainly can't explain how science denial came about. The script is too intricate to have happened by accident. The explanation seems more likely one of *malfeasance*.

The five tropes form a strategy that was deliberately created by those who had an interest in getting people to deny selected findings of science that threatened their beliefs. This was then copied in subsequent campaigns and used against different scientific findings, until it is now a battle plan that can be used to "fight the science" on just about any topic. Science denial is not an error, it is a lie. Disinformation is intentionally created.[40]

In their important book *Merchants of Doubt*, Naomi Oreskes and Erik Conway tell the story of how cigarette companies in the 1950s were hair-on-fire worried about a scientific study that was about to be published, which showed an all-but-causal link between cigarette smoking and lung cancer.[41] Instead of continuing to fight among themselves over whose cigarettes were "healthier," the executives of the largest tobacco companies joined forces and hired a publicist to devise a strategy. Fight the science, he advised. Manufacture doubt. Create as many reasons as you can for thinking that the scientists might be biased and telling only one side of the story. Then get your side out. Hire your own experts. Come up with your own "scientific" findings. Take out full-page ads in the popular press to call the scientists' findings into question. Insist that any alleged link between smoking and lung cancer must be *proven*.[42]

Sound familiar?

Oreskes and Conway do a superb job of showing how the cigarette companies went on to create a campaign of disinformation where—in the words of an infamous 1969 memo written by a tobacco executive—"doubt is our product since it is the best means of competing with the 'body of fact' that exists in the mind of the general public. It is also the means of establishing controversy."[43] This enabled the tobacco companies to bamboozle the American public over the ensuing decades as they held out for "proof" while continuing to profit by selling cigarettes. Unfortunately, this campaign was so successful that it became the blueprint for all subsequent science denial—the authors call it the "tobacco strategy"—which was then used on acid rain, the ozone hole, climate change, and so on.[44]

With smoking and cancer, denial was created to serve an obvious corporate interest.[45] With climate change, it looks to have been the same.[46] For more on this, I highly recommend Oreskes and Conway's book. My aim is not to give the complete history of science denial, but to see if we can learn how to talk to a science denier in order to get them to change their mind. Obviously, this will not work with someone who has cynically created a lie (whether or not they actually believe it). I therefore propose that we move on to others—the audience for science denial—to assess their reasons for believing something that they did not invent and may derive no obvious benefit from believing to be true.[47]

Here it is important to realize that there are many possible motivations for science denial. Economic self-interest is an obvious one. But there are also political, ideological, or religious reasons why someone might prefer to deny a particular scientific finding, which might be deeply personal. When these interests are exploited by those who have created a campaign of denial, they might attract millions of followers to do their bidding. Ignorance and gullibility may be a factor. But there has to be something else at work too. Whether or not climate denial was created because it serves someone else's economic interests, the question remains: why does the science denier actually believe it?

At times, personal interests for the believer may be at stake. Even if these are not economic, they can be powerful. A smoker might have had a built-in reason to welcome the news that there was "another side" to the scientific findings on tobacco in the 1950s. Motivated reasoning is a powerful psychological force, whereby we are more prone to look for information that backs up the things we *want* to believe, as opposed to facts that might cause us psychic discomfort. If a person did not want to stop smoking, for instance, wouldn't it be better to believe that it was safe? People can confabulate all sorts of information or lie to themselves when they want to. Research has shown that for the most part, this is not even done at a conscious level.[48] Perhaps this is why there is such a close kinship in our minds between someone who is "in denial" and someone who is a "denier."[49] We lie to ourselves as a means of more efficiently lying to others.

Seventy years of social psychology has shown that satisfying the human ego is an important part of our behavior. And an integral part of this is maintaining a positive view of ourselves. This can account for those times when we resolve cognitive dissonance by telling ourselves a story we would prefer to believe over one that is true, as long as we can be the hero in it. It also involves making sure that we present a favorable image of ourselves to those in our social circle, whose opinions we care about. Thus, our beliefs and behavior are formed in a hothouse of self-opinion, as reflected back to us in the opinion of others. Is it any surprise that our beliefs on empirical topics are thereby based on more than just facts, but also the psychological and motivational forces that shape all behavior and beliefs? As such, our empirical beliefs are ripe for manipulation, whether from our own interests or someone else's.

One also should not underestimate the simple role of fear. Cutting-edge neuroscience, using fMRI scanners, has shown that, when exposed to ideas that threaten their beliefs, conservative partisans experience more activity in the amygdala (fear-based) part of their brain than liberals do.[50] Might the same prove true for science deniers?[51] When a new parent hears that vaccines may be dangerous to their baby, how could this not make them anxious? As they turn to Google and come upon some alarming misinformation, their brain floods with cortisol. When they then turn to their family physician and get dismissed—"I can't believe you'd give any credence to that bunk"—they may feel disrespected, and so turn to an anti-vaxx conference for more information. By that point it is too late. As one journalist who got kicked out of an anti-vaccine convention put it:

> "AutismOne—and the anti-vaccine world as a whole—works remarkably well as an engine for radicalization. Parents are brought in with a genuine concern for their children's health, and a desperation to find answers, and are met with a variety of new and increasingly wild claims about the medical establishment, the government, and, ultimately, the secret rulers of the world."[52]

Another psychological force behind science denial might be a feeling of alienation and disenfranchisement. Of course, rude treatment, name-calling, and dismissal as an idiot by those who are attempting to fight science deniers can itself cause a sense of alienation.[53] But I am talking about something deeper than that. When I was at FEIC 2018, I noted a disproportionate number of people who had had some sort of trauma in their lives. Sometimes this was health-related, other times it was interpersonal. Often it was unspecified. But in every instance the Flat Earther referred to it as in some way related to how they "woke up" and realized that they were being lied to. Many of them embraced a sense of victimization, even before they became Flat Earthers. I have found very little in the psychological literature about this, but I remain convinced that there is something to learn from this hypothesis.[54] I came away from the convention with the feeling that many of the Flat Earthers were broken people. Could that be true for other science deniers as well?

Whether this is right or not, it seems clear to me, based on both my personal experience and study of the literature, that most science deniers embrace a heady stream of resentment and anger at "elites" or "experts" who purport to tell them the truth. Some of this is related to Tom Nichols's earlier-cited work, *The Death of Expertise*, where he explores the kind of populist grievance that has led to so much of our post-truth culture. This goes wider than science denial. Indeed, even though I believe that science denial was one of the roots of post-truth,[55] this has now doubled back on itself in support of an entire denier culture, involving everything from climate change to vaccinations to wearing masks during a pandemic, that has made science denial worse. As familiar battle lines are redrawn—sometimes along partisan lines—it becomes easier to feel alienated. With separate information sources, fragmentation, polarization, and the creation of an "us against them" mentality, it should come as no surprise that science would get caught up in the post-truth vortex.

Is this to say that science denial is now simply political? In part, this may be true. The obvious example here is climate change denial, which suffers from a 96 to 53 percent partisan split.[56] In his work on science denial, cognitive scientist Stephan Lewandowsky argues that these days, virtually all science denial trends conservative:

> There has been a decades-long, gradual erosion of trust in the scientific community among conservatives—but not liberals—since the 1970s. . . . This erosion of trust has coincided with the emergence of multiple scientific findings that challenge core conservative views, such as belief in the importance and beneficence of unregulated free markets. . . . In summary, the rejection of specific scientific evidence across a range of issues, as well as generalized distrust in science, appears to be concentrated primarily among the political right.[57]

Even Lewandowsky admits, however, that the underlying cognitive forces that lead to such things as belief in conspiracy theories or confirmation bias are not limited to conservatives. We all have the same brains and cognitive biases that were shaped by the same evolutionary forces.[58] Thus, the question remains of whether it is possible for there to be instances of "liberal" science denial, which I will consider in chapters 6 and 7.

The political angle notwithstanding, we now stand on the doorstep of a key insight into the question of why science deniers believe what they believe, even in the face of contravening evidence. The answer is found in realizing that the central issue at play in belief formation—even about empirical topics—may not be *evidence* but *identity*.

Identity can be found within a political context, but that is surely not the only place it exists. People can find a sense of identity within their church, their school, their family, their profession, their neighborhood, or ultimately within their peer group of fellow science deniers. In his brilliant book *Know-It-All-Society*, Michael Lynch gives an account of how our beliefs become our convictions, and the relationship this bears to identity.

> A conviction is a belief that takes on the mantle of commitment—a call to action—because it reflects our self-identity. It reflects the kind of person we aspire to be, and the kinds of groups and tribes we wish to belong to. That is why attacks on our convictions seem like attacks on our identity—because they are. But that is also why we often ignore evidence against our convictions; to give them up would be to change who we imagine ourselves to be.[59]

The psychological roots of this can be found in what Yale scholar Dan Kahan calls "identity-protective cognition."[60] We might think that making up one's mind on a scientific topic is a simple matter of looking at the data. And, in fact, when we are concerned with a topic where the results do not tread on one of our sacred convictions, that is usually the case. Remember the idea that science deniers are cafeteria skeptics? Even a denier can usually sort out the right answer to a scientific question based on the data, as long as it does not involve a belief that is central to their self-identity. But when we begin to deal with "controversial" subjects like evolution or climate change (or, for some, the shape of the Earth), our reasoning skills go out the window. Not only do we not change our mind, we can't even do a rational assessment of the evidence.

The tension here is between what Kahan calls the "science comprehension thesis" and identity-protective cognition. The SCT model is based on the idea that the best way to convince someone that an empirical hypothesis is true is to give them enough information to make a rational decision about it. Treat them like a scientist. If one is rational, and understands how to reason based on evidence, it should be fairly straightforward to judge whether a conclusion is warranted. Here we are assuming that the only reason someone would reject a well-supported scientific theory is that they are irrational (or stupid, or unskilled) or that they do not have sufficient information. A better name for this, I think, is the "information deficit model,"[61] because we are assuming that any instance of science denial can be remedied by giving the science denier more information. How often have we seen scientists try this! When the climate change denier says that there has been no global warming since 1998, we give them more temperature data. When they doubt this, we might move on to the loss of sea ice. When they doubt that, we move to something else. Ultimately, we may assume that they are irrational and simply walk away. If you cannot convince someone with evidence, why continue talking to them at all? But what if the problem is *not* that they have insufficient evidence? What if the hold-up is that they are engaging in identity-protective cognition?

To test this, Kahan performed an experiment on assessing the effectiveness of a new (fictitious) skin cream. To my knowledge, there is not and never has been any sort of science denial (or identity-based reasoning) on the basis of skin cream. Kahan set up the experiment with one thousand experimental subjects, who were first surveyed to assess their political beliefs, then given a set of fabricated data (see figure 2.1).[62]

After doing a little math, we have all of the information we need to see whether the skin cream is effective on a rash. At first glance, it may appear that the cream was effective. After all, 223 of the people who used it saw their rash get better, whereas only 107 of the people who did not use the cream saw any improvement. But the numbers must be leavened by considering those whose rash got *worse*. Contrary to one's initial impression, the appropriate conclusion to draw is that the skin cream is *not effective*. After all, 25 percent of the people who used the skin cream saw their rash get worse, compared to only 16 percent of the people who did not use it.[63]

Kahan found that most people were terrible at coming to the right conclusion. But the results did not break down along party lines! Instead, perhaps predictably, the only difference in answers was between people who were good at math and those who weren't, which was perfectly consistent with the science comprehension thesis.

But then Kahan did another version of the test, *using the exact same data*, on an ideologically charged topic: whether gun control increases or decreases

Results

	Rash got better	Rash got worse
Users of new skin cream	223	75
Nonusers of new skin cream	107	21

Figure 2.1
Fabricated data used as stimuli in an experiment by Kahan et al., "Motivated Numeracy and Enlightened Self-Government" (2013).

Results		Results	
Increase in crime	**Decrease in crime**	**Decrease in crime**	**Increase in crime**

Cities that banned guns

| 223 | 75 | 223 | 75 |

Cities that did *not* ban guns

| 107 | 21 | 107 | 21 |

Figure 2.2
Fabricated data used as stimuli in an experiment by Kahan et al., "Motivated Numeracy and Enlightened Self-Government" (2013).

crime (see figure 2.2).[64] In the first iteration (left), the data showed that gun control was correlated with a *decrease* in crime. In the second (right), it showed an *increase*.

Here, the results were different. As political writer Ezra Klein explains in his account of the experiment:

> Presented with this problem a funny thing happened: how good subjects were at math stopped predicting how well they did on the test. Now it was ideology that drove the answers. Liberals were extremely good at solving the problem when doing so proved that gun-control legislation reduced crime. But when presented with the version of the problem that suggested gun control had failed, their math skills stopped mattering. They tended to get the problem wrong no matter how good they were at math. Conservatives exhibited the same pattern—in reverse. . . . Being better at math didn't just fail to help partisans converge on the right answer. It actually drove them further apart. Partisans with weak math skills were 25 percentage points likelier to get the answer right when it fit their ideology. Partisans with strong math skills were 45 percentage points likelier to get the answer right when it fit their ideology. . . . People weren't reasoning to get the right answer; they were reasoning to get the answer they wanted to be right.[65]

One might conclude from this that politics exacerbates a flaw in human reasoning about empirical topics. But that is perhaps too fine a point to draw about science denial. Yes, the reasoning and beliefs in Kahan's experiment were framed within a political context, but politics is only one way of forming identity. What if the underlying issue here is not merely that *politics* can interfere with our reasoning skills but that *all identity* can? Indeed perhaps identity is more important than any specific ideology. It is after all called *identity*-protective cognition.

In her important paper "Ideologues without Issues: The Polarizing Consequences of Ideological Identities," Lilliana Mason argues that the driving factor behind political polarization is not any of the "issues" we might think of as typically liberal or conservative but instead the mere fact of having a partisan label that gives one an identity.[66] The important thing is to pick a team, so that we know which side to root for in the political game of us against them.

In her research, based on survey data, Mason found that the strength of a person's affiliation with a political identity was a much greater predictor of how they felt about the "other side" than the ideological content that might lie behind the identity. Subjects were surveyed on their opinions on six issues: immigration, gun control, same-sex marriage, abortion, Obamacare, and the deficit. Then they were asked how they felt about marrying someone of the opposite political affiliation. Or being friends with them. Or simply spending time with them. What Mason found was that the identity-based label had twice as much predictive value for how subjects felt about the "other side" as did their opinion on any of the six issues![67] Conservatives were actually much more moderate in their political views than liberals, but no less partisan in their identity. Mason marked this overall difference as one between "identity-based ideology" versus "issue-based ideology."[68]

But if it is *identity* that partisans care about more than the content of their ideology, this might make us wonder in what sense identity-based reasoning is "ideological" at all. In his essay "People Don't Vote for What They Want. They Vote for Who They Are," philosopher Kwame Anthony Appiah notes that during the Trump era, Republicans have turned nearly 180 degrees from their previous position on Russia. The photo accompanying Appiah's article shows two Trump supporters at a rally wearing T-shirts that say "I'd rather be a Russian than a Democrat." In concert with Mason's reasoning, Appiah concludes that "identity precedes ideology."[69]

Is it possible that the *content* of science denialist beliefs is similarly superfluous, or at least malleable? What if the Flat Earthers I spoke to at FEIC 2018 were motivated to hold their beliefs *not* because it made sense to them, but because it plugged some hole in their psyche? It gave them a team to root for and fed their sense of grievance. Perhaps it also made them feel better about their situation in being alienated from society and its "normie" beliefs, because they were now connected to a group of people who agreed with them and told them that they were right. If someone wants to fit in, perhaps the content of their belief merely comes along for the ride.[70] Is this perhaps why it is so hard to change a science denier's mind with evidence, because in some sense the evidence isn't really what their beliefs are about? The content of the belief may not be as important as the social identity it affords.

There are powerful cognitive forces that seduce us into believing what we want to believe. What the people around us—whom we know and trust— want us to believe.[71] And these days, when we can find a whole community of others who agree with us, it's much easier to hold fringe beliefs. Either online or in person, when you are in a crowd it is easy to pick a side and demonize those who disagree with you. Once you decide *who* to believe, perhaps you know *what* to believe. But this makes us ripe for manipulation and exploitation by others.

Perhaps this provides the long-awaited link between those who create the disinformation of science denial and those who merely believe it. If some person or organization with a powerful agenda has an interest that conflicts with a scientific finding, it's not hard to stir the pot of partisanship or polarization along "identitarian" lines to get someone to come over to your way of thinking. This seems to be exactly what happened with corporate denial over the link between smoking and lung cancer that started in the 1950s. And it happened later, in a confluence of corporate and political interests, over climate change as well. In this way, special interests were able to create a sense of identity around issues from which believers got no material benefit. But does this mean that *all* science denial is the result of outside interests? This is much harder to prove. While religious ideology seems to be firmly behind the creationist/intelligent designer's pushback against Darwinian evolution, what are the corporate/ideological interests behind Flat Earth? Or anti-vaxx? Or GMO denial? Short of inventing a conspiracy theory, I can't see that any exist.

Sometimes erroneous beliefs happen organically, due to factors too diverse to categorize, and may end up creating an identity or interest group de novo. When that happens, for whatever reason, others jump aboard. After all, perhaps it is the team, not the ideology. We want to be on someone's side. And it is important to remember that no matter how it gets created, we must fight science denial by talking to the believers, not the cynical folks who might have thought it up. While it can be helpful to try to expose a campaign of disinformation, this is not the primary way to overcome science denial. Once the lies are out there, even if they are exposed they have already done their damage. We must talk to the people who believe them. If one can uncover corruption and malfeasance, that's all to the good. But even if that doesn't exist, or despite it, we still need a means to fight back.

Whether the lies are manufactured by cynical outside interests or confabulated from our own psychic wounds or ego, we arrive at the same place. Science denial is *not* based on lack of evidence. Which means that it cannot be remedied just by providing more facts. Those who wish to change the minds of science deniers have to stop treating them as if they were just misinformed colleagues who already know how to reason about evidence but just don't have the data. No amount of evidence is ever going to change the mind of a science denier, if we do not appreciate the role that their beliefs play in reinforcing their social identity.

In their tremendously helpful book *How to Have Impossible Conversations*, philosopher Peter Boghossian and mathematician James Lindsay give us this surprising advice in trying to convince someone who disagrees with us: avoid facts!

> The most difficult thing to accept for people who work hard at forming their beliefs on the basis of evidence is that not everyone forms their beliefs in that way. The mistake made by people who form their beliefs on the evidence is thinking that if the person with whom they're speaking just had a certain piece of evidence then they wouldn't believe what they do.[72]

Instead, the authors advise us to ask disconfirming questions like, "What facts or evidence would change your mind?"[73] This is precisely what I did at FEIC 2018, even though I hadn't read Boghossian and Lindsay's book yet. (That's likely because—as fellow philosophers—we both borrowed the strategy from Karl Popper.)[74]

At this point the five tropes of science denial come back into play. We have already seen that they form a common script that lies behind all science denial reasoning. Why is that important? *Because once you know the script, you can challenge it.* The five-tropes script allows science deniers to feel that they are actually reasoning, rather than just reinforcing what they are motivated to believe on the basis of their identity. I am not saying that they learn this script verbatim, or even realize its existence, but they do internalize its elements and can get pretty good at using them. But if you can disrupt the script, you might have a fighting chance of convincing them. Get them to question the talking points that have been provided by their group. For a moment, get them to think for themselves. The goal of talking to a science denier is to create an opportunity for doubt, where you might get them to see things from a different point of view.

Of course, it is next to impossible to get someone to change their beliefs against their will. No matter how good you are at rhetoric (or philosophy), you aren't going to catch a science denier in a logical contradiction and get them to change their mind. Remember that when you challenge some-one's beliefs, you are challenging their identity![75] This is not to say that you cannot use empirical evidence to convince someone; just remember that evidence is a tool within a larger conversation whose goal is to try to get the science denier to try on a new identity. To see what it feels like to care more about evidence. To gain a greater appreciation for what it means to think like a scientist.

When challenging their script and working in your own evidence, always be cognizant of the *real* reason that science deniers believe what they do: because of how it makes them feel. This means that you have to take into account not just the science denier's beliefs, but how they justify those beliefs. The script is how they *defend* their beliefs, but it is not why they *have* them. They have them to resolve their fear, or to feel less alien-ated, or to embrace a coveted social identity. What they believe is a reflec-tion of who they are.[76]

In broadest scope, science denial is an attack not just on the content of certain scientific theories but on the values and methods that scientists use to come up with those theories in the first place. In some sense, science deniers are challenging the *scientist's* identity! Science deniers are not just ignorant of the facts but also of the scientific way of thinking. To remedy this, we must do more than present deniers with the evidence; we must get them to rethink how they are *reasoning* about the evidence. We must invite them to try out a new identity, based on a different set of values.[77]

I'm afraid this means that we must abandon the information deficit model once and for all. You cannot convert a science denier simply by filling in the knowledge they lack. Again, this does not mean that facts don't mat-ter—or that there is no role for evidence—but it matters how this evidence is presented. And by whom. And the epistemological context within which they are receiving it. Time and again at FEIC 2018, Flat Earthers rejected my evidence because they did not trust the scientists who had created it. The deficit they experienced was not one of information, but one of *trust*. You won't change somebody's deeply embedded beliefs—what Michael Lynch calls their convictions—simply by giving them new facts or even a new way of thinking. You have to help them deal with the threat that new information poses to their identity.

3 How Do You Change Someone's Mind?

Now that we understand a bit more about the elements of science denial—and some of the causes and motivations that lie behind it—the question naturally arises of what we might be able to do about it. At this point it would be criminal not to review the empirical literature, which suggests that, under some circumstances, people *can* be convinced on the basis of evidence. As we'll see, though, the answers aren't quite definitive, which means that we'll have to turn to the anecdotal literature too, which is wonderfully illuminating.

Changing Minds within an Experimental Setting

In August 2000, James Kuklinski and colleagues published a paper called "Misinformation and the Currency of Democratic Citizenship," in which they studied how partisans can change their minds. Although they assessed a political rather than scientific topic, the point is still relevant for our purposes, for a key part of getting subjects to alter their beliefs had to do with empirical evidence. The topic of the study was welfare. Researchers wanted to test subjects' knowledge about welfare, then see whether it was possible to get them to change their views when presented with correct information.[1] As expected, subjects' knowledge of welfare was atrocious, with only 3 percent of the surveyed sample able to recount even half of the correct facts about the average American welfare payment, the percentage of welfare recipients who were African American, and what percentage welfare constituted of the total federal budget. It was clear that subjects were not merely uninformed but misinformed. Researchers noted, moreover, a perverse effect (that has since been confirmed in other research): subjects who were the least informed tended to be the most confident that their views were correct.[2]

In a telephone survey of 1,160 Illinois residents, Kuklinski and his colleagues gathered data to measure the extent of subjects' mistaken beliefs about welfare.[3] In a follow-up survey, they zeroed in on the single question of what percentage of the federal budget was taken up by welfare payments, but then added a question concerning what level respondents thought *should* be taken up by welfare. This was done as a way to prime subjects to think about their attitudes toward welfare juxtaposed against their (mistaken) beliefs. It also gave experimenters a way to measure subjects' tacit support for welfare (on the theory that a larger gap between what they thought was the actual level and what they thought was an ideal level indicated less support). Then researchers tried something bold: after asking subjects to answer the two prior questions, they provided half of them with correct information (leaving the other half as a control), then overtly asked both groups whether they supported welfare spending.

The results were noteworthy. Since all subjects were so highly misinformed, it was common for them to wildly overestimate federal spending on welfare. A typical response was to say, for instance, that 22 percent of the federal budget went to welfare, followed by a statement that only 5 percent *should* go to it. But for those subjects who were then hit between the eyes with the correct figure that only 1 percent of the federal budget actually went to welfare, *a statistically significant number of them then indicated levels of support for welfare that were deeply at odds with the attitude one would have predicted given their previous answers.* After being exposed to accurate information, they softened their views. In the control group, no difference was noted between subjects' predicted and actual levels of support.[4]

Kuklinski et al. write:

> Respondents take notice when told that the percentage actually spent on welfare is even lower than their preferred level. Misinformed citizens, then, do not always remain oblivious to correct information. If it is presented in a way that "hits them between the eyes"—by drawing attention to its policy relevance and explicitly correcting misperceptions—such information can have a substantial effect.[5]

In a 2010 study entitled "The Affective Tipping Point: Do Motivated Reasoners Ever 'Get It'?" David Redlawsk and colleagues hypothesized that—unless no one ever actually changes their mind about anything—even motivated reasoners must reach a "tipping point" where they begin to be swayed by information that is at odds with their beliefs.[6] Once again, this experiment was done with political (rather than scientific) beliefs—in this case measuring commitment to a preferred political candidate. This time, however, the experiment was done in person rather than by telephone, with 207 nonstudent subjects in eastern Iowa, in a mock political campaign.

Before they began, researchers noted prior findings that suggested that *all* beliefs are formed based on not just informational factors but also affective ones. This is to observe that, when presented with information that is incongruent with our beliefs, we will react not just to its content but also to how we *feel* about it.[7] In the case of commitment to a political candidate (which is what this study measured), Redlawsk et al. found that if subjects were already committed to a candidate, then a small amount of negative information actually *increased* their level of support. This may sound irrational (and it may well be), but it is part and parcel of one of the ideas we have already encountered called motivated reasoning, which is that we do not simply gather information passively but factor in our feelings, especially in deciding whether to change our beliefs. Remember the partisans in Kahan's study, who were eager to find facts that supported their preexisting ideological commitments about guns and crime? The preservation of one's identity, and the reduction of any cognitive dissonance that would threaten it, are two of the most foundational ideas in social psychology. It should come as no surprise, then, that how we *feel* about a belief would affect whether we are willing to hold it. As we have seen, if we propose to study whether facts are capable of changing minds, we had better be ready to consider the social and emotional context in which those facts are presented.[8] On the whole, people will defend even factually incorrect beliefs that are congruent with how they want to view themselves.

Unless we are prepared to go on this way ad infinitum, however, Redlawsk and colleagues wondered at what point we become ready to update our beliefs on the basis of negative information. When we encounter information that is at odds with our preferred beliefs, it may cause us to hold those beliefs more strongly for a while, but anxiety eventually increases to the point where something has to give. If the negative information is bad enough—and repeated often enough—researchers hypothesized that we would eventually reach a tipping point at which we would accommodate the disconfirming information and update our beliefs.

In this particular experiment, researchers held a mock political primary, with made-up candidates, and controlled the flow of positive and negative information about them, interspersed with polls asking how subjects felt about their choice. They found that no matter how committed a subject might be to a given candidate, if there was enough negative information they eventually reached a point where they would abandon their choice. That point may have been different for each subject, but everyone *had* a tipping point. The researchers write, "At some point our voters appear to wise up, recognize that they are possibly wrong, and begin making adjustments. In short, they begin to act as rational updating processes would require."[9]

Of course, even though this study used live subjects in a face-to-face setting, it was still within an experimental, simulated environment. This raises the question of whether we can expect that everyone would react similarly, in the real world with real-life candidates.

> It is easy to imagine a long-time fan of a presidential candidate rejecting virtually all new negative information about him or her and sticking to an early evaluation. Yet even such a fan might, in the face of overwhelming information counter to expectations, awake to the changed reality and revise her beliefs accordingly.[10]

The analogy between political candidates and empirical beliefs is a seductive one, where we might imagine that subjects who previously had been attached to a particular belief—say, that the Earth was flat—would in the right circumstances give it up. Does this suggest that—in the face of an onslaught of repeated negative information—even science deniers might change their minds? If we hit them right between the eyes over and over with negative or disconfirming facts, would they eventually begin to update their beliefs in a rational way?

Perhaps, but we don't know yet. The problem is that neither the Kuklinski nor Redlawsk studies deal with explicitly scientific beliefs. While their results are suggestive, they do not provide direct empirical support for the idea that facts and evidence can be used to overcome science denial. Still, it is encouraging to note their conclusion: of course facts matter! Otherwise, why would anyone ever change their mind about anything?

But all of this optimism was thrown into doubt by a landmark 2010 study published by Brendan Nyhan and Jason Reifler entitled "When Corrections Fail: The Persistence of Political Misperceptions," which proposed something popularly known as the "backfire effect."[11] Here, the experimental design involved exposing political partisans to corrective information that would challenge their false beliefs. For conservatives, it was the idea that Iraq had weapons of mass destruction (WMDs) before the Iraq War. For liberals, it was the idea that President George W. Bush had imposed a total ban on all stem cell research. Both of these claims are factually false.

All participants were initially primed with bogus newspaper articles that seemed to corroborate these false beliefs. This was done to provide both liberals and conservatives with reason to think that the claims were true. But when later presented with credible corrective information—for instance, a quotation from a speech that Bush had given, in which he admitted that there had been no WMDs in Iraq—results broke down upon partisan lines. Perhaps unsurprisingly, liberals and moderates accepted the corrective information and changed their beliefs. Conservatives did not. In fact, researchers noted that some conservatives actually became *more* convinced that their original (mistaken) belief was true, once they had been presented with correcting information. This perverse phenomenon was deemed the "backfire effect," whereby some partisans not only refused to give up their original beliefs *but held onto them more strongly* once they were challenged.

When the time came to provide corrective information to the "liberal" belief—that Bush had imposed a total ban on all stem cell research (whereas in fact he had imposed a limited ban only on federally funded research on stem cell lines created before August 2001, with no ban on private research)— the correction worked for conservatives and moderates, but not for liberals. Notably in this case, however, researchers found no backfire effect. Although the corrective information did not convince liberals to give up their mistaken belief, it did not cause them to hold it more strongly. This is to say that in both conservative and liberal cases, the corrective information failed to convince partisans, but only for conservatives did it cause them to double down on their mistaken belief.

These findings knocked the fact-checking community on its ear. After the 2016 presidential election, things devolved into a full-blown panic, with several splashy headlines like the aforementioned "This Article Won't Change Your Mind" (*Atlantic*) and "Why Facts Don't Change Our Minds" (*New Yorker*) that led people to conclude not only that you probably couldn't change people's minds with facts, but that in trying to do so you might make things worse.[12] In light of this, might it be a mistake to think that we can (or should) try to challenge the beliefs of science deniers?

Then in 2017, the whole thing got revised again, when Ethan Porter and Thomas Wood found that the backfire effect could not be replicated.[13] It is important to be clear that Porter and Wood did *not* undercut the major finding of Nyhan and Reifler's study. Partisan subjects still resisted factual information, and found it difficult to change their minds on its basis. But the backfire effect had disappeared. Researchers speculated that perhaps this meant the effect was a unicorn that didn't show up very often. Nyhan and Reifler chimed in to point out that in their original study, the backfire finding was only a very small part of their overall results, and existed only in limited circumstances for a few ultra-partisan subjects. All four researchers agreed that the majority of subjects still did not change their minds on the basis of corrective information. As one commentator put it, "We are fact resistant, but not fact immune."[14]

In a nod to scientific openness and integrity, Nyhan and Reifler joined Porter and Wood in sharing and promoting this result. When presented with a challenge to their original finding, they cooperated fully in its revision.[15] Of course, word didn't immediately trickle out to the general public, and the perception that we might be doing more harm than good by trying to change people's minds with facts still lingers. But for the scientific community, the cloud had now lifted. Further work was possible on how best to change minds.

Indeed, one of the most intriguing new studies was done by Nyhan and Reifler themselves. In a 2017 paper entitled "The Roles of Information Deficits and Identity Threat in the Prevalence of Misperceptions," Nyhan and Reifler took up two of the thorniest questions we have encountered so far: whether mistaken beliefs about empirical matters can be corrected through making up the information deficit of the person who holds it, and whether this depends at all on considerations about potential threats to the "identity" (self-esteem, self-concept) of the person who holds the beliefs.[16]

This new study sought to answer two specific questions. One was whether the manner of presentation of correcting information mattered in overcoming the information deficit that was presumably behind subjects' mistaken beliefs. The other was whether subjects' resistance to corrective information might be mitigated if one could improve their self-opinion. Here again, the topic of concern was changing political beliefs, but in *one* of the three experiments the "political" belief chosen concerned whether subjects were willing to accept corrective information about the reality of global warming. Finally, a connection to science denial! In this trial, subjects had already been screened for their predilection to find such corrective information threatening: they were all Republicans. Information about the truth of climate change would thus be in direct conflict with their identity.

Nyhan and Reifler found that the form in which corrective information was presented had a statistically significant effect. Graphs worked better than text. In fact, graphs alone were so effective that they were not enhanced by providing textual support. Unfortunately, no effort was made to probe the question of why graphs were so effective. Perhaps they seemed more objective? Perhaps they presented less opportunity for the sort of rhetoric or divisive language that can threaten one's ego?

Researchers also measured whether it made a difference how subjects *felt* when they were receiving corrective information, graphical or otherwise. Based on their hypothesis concerning identity threat, researchers conjectured that if subjects were confronted with a belief that threatened their self-opinion, they would be more likely to reject it; therefore, if they could somehow be made to feel less threatened, perhaps they might be more willing to accept corrective information. To reduce the likelihood of identity threat, Nyhan and Reifler gave subjects an exercise in self-affirmation immediately before presenting corrective information. In short, they tried to make subjects feel good about themselves. The results were somewhat equivocal. While they found some effect in certain circumstances (depending on how strongly subjects identified with the values of the Republican Party), these results were dwarfed by the form of presentation. Graphs ruled. Although Nyhan and Reifler found a positive effect for self-affirmation, it was a weak one.

Here it is tempting to try to find fault with researchers' methodology. Even if their hypothesis about ego or identity threat was correct, why suppose that this could be mitigated by self-affirmation? Citing prior studies on precisely this question, researchers noted that they were "disappointed" by their results. Others, however, might understand that their findings were completely predictable. Are we really to believe that partisan Republicans— who may have been fed misinformation and told over and over that the rejection of climate change is part of the core identity of their party—are supposed to overturn their beliefs because they have engaged in a simple exercise to make them feel better about themselves? This stretches credulity. In fact, perhaps the best evidence for the hypothesis about the importance of identity threat was right under researchers' noses—for what can be more neutral and less confrontational than lines or bars on a graph? Perhaps the mode of presentation *was* the social context. In the end, Nyhan and Reifler observe that "these results suggest that misperceptions are caused by a lack of information as well as psychological threat, but . . . these factors may interact in ways that are not yet well understood."[17]

A great opportunity for further empirical research presents itself here. Nyhan and Reifler are likely right that belief formation and change are not just a matter of having correct factual information, but of the emotional, social, and psychological context within which beliefs are formed. As we saw in Kahan and Mason's earlier experiments, identity might make a crucial difference. The very fact that such an otherwise dry scientific issue as whether the Earth is warming has been made into a matter of partisan disagreement reflects the power of politics to shape empirical beliefs. As we all know, with the right motivation (and misinformation), even factual matters can be polarized. As a result, some truths can be made to threaten our identity or membership in a particular group.[18] It thus matters a great deal not just what the facts are but how they are presented and by whom. Can I trust the source? Do they have a political interest in showing that I am wrong? As we have seen, even scientific beliefs can be manufactured into a matter of identity. If we are Republicans, does the truth about climate change threaten us? Likely yes. But if we are trying to get someone to change their mind on a factual topic that is shot through with such partisan or other ideological identity, how best to approach them?

It is important to remember our common sense. If we are trying to get someone to change their mind on any subject, a personal touch can do wonders. And here too the manner of presentation surely matters. Should we yell at them or call them names? Insult their intelligence? Probably not. It is much more effective to approach such conversations in a nonthreatening manner. We should seek to build trust, show respect, listen, and remain calm. It just makes sense that being hostile toward someone we disagree with would get their back up. Since belief formation has been shown to be an amalgam of both information and affect, why would belief change be any different? Yet the effectiveness of such common-sense tactics are ill-studied in the experimental literature, particularly on the subject of science denial.

In his previously cited essay "How to Convince Someone When Facts Fail," Michael Shermer offers common sense to compensate for the experimental void. He begins in the same place as Nyhan and Reifler: "people seem to double down on their beliefs in the teeth of overwhelming evidence against them. The reason is related to the worldview perceived to be under threat by the conflicting data."[19] The evidential basis for this in the literature is not in question, and in fact goes back to Leon Festinger's classic finding on cognitive dissonance.[20] When subjects are sufficiently motivated—and their ego or identity is threatened—they will resist all efforts to get them to concede that they are wrong. After acknowledging the classic backfire effect, Shermer offers important advice—already quoted in the introduction to this book—about how to convince people: keep emotions out of it, don't attack, listen carefully, and always show respect.[21]

Since Shermer is a professional skeptic with decades of experience—who encounters science deniers in the wild as a matter of course—we would be wise to heed his advice. Though the experimental literature may not yet corroborate the efficacy of these tactics in practice, they are surely upheld by looking at how people actually change their minds in the real world. So now that the backfire effect has been overthrown, how much longer should we wait? Global temperatures continue to increase. Anti-vaxxers threaten that even once there is a vaccine for COVID-19, they might not take it. So why not get out there and try to change some minds?

It was at this point that I went on the road and tried to embrace Shermer's advice in a series of face-to-face encounters with science deniers. If it is true that an important part of belief change involves dealing with identity issues, perhaps it is better to go outside the laboratory and deal with this in person. We can provide corrective information online, by phone, or in a simulated mock-up experimental setting, but we can best build trust face-to-face. Perhaps, under just the right circumstances, we might change someone's mind as part of an experiment, as Kuklinski and Redlawsk have shown. But could this be done in the wild and, specifically, on science denial topics? That's what I went to FEIC 2018 to find out.

Breakthrough

In the summer of 2019—seven months after I got back from talking to Flat Earthers—a truly groundbreaking study by Cornelia Betsch and Philipp Schmid came out in *Nature Human Behaviour*, which provided the first direct empirical evidence that you *could* change the minds of science deniers.[22]

They even provided a script for what to say. I am exaggerating only slightly when I say that I could have read this paper even if my hair were on fire. Although I already mentioned this study in the introduction, it is worth reviewing their results in a little more detail, now that we are prepared to start using them.

Schmid and Betsch ran six online experiments with 1,773 subjects in the United States and Germany on topics such as climate change and vaccine denial. And what they found was astonishing. Fighting back against science denial not only had a positive effect on changing subjects' beliefs, the effect was greatest for subgroups who had the most conservative ideologies. Like Porter and Wood, they found no backfire effect. In the course of their work, Schmid and Betsch tested four possible ways of responding to subjects who had been exposed to scientific misinformation: no response, topic rebuttal, technique rebuttal, and both kinds of rebuttal. Topic rebuttal consisted of providing subjects with information to correct the bogus content of a message they had just heard. For instance, if subjects had encountered a claim that vaccines were unsafe, this might be opposed by highlighting the excellent safety record of vaccines.[23] A second strategy was to go back and borrow a page from the earlier discovery—already discussed in chapter 2—that there is a common script for virtually *all* science deniers. Schmid and Betsch call this technique rebuttal, and it involves pointing out the five dangerous techniques—cherry-picking, reliance on conspiracy theories, use of fake experts, illogical reasoning, and reliance on impossible standards for scientific reasoning—to mitigate their impact on subjects' potentially denialist beliefs. For instance, in response to the same message about how vaccines were unsafe, experimenters might respond by pointing out that it is unreasonable to hold vaccines to the standard that they must be 100 percent safe, since no medication—not even aspirin—can meet this.[24]

The clear result of this study was that providing no response to misinformation was the worst thing you could do; with no rebuttal message, subjects were more likely to be swayed toward false beliefs. In a more encouraging result, researchers found that it was possible to mitigate the effects of scientific misinformation by using either content rebuttal or technique rebuttal, and that both were equally effective. There was, moreover, no additive advantage; when both content and technique rebuttal were used together, the result was the same. This means that advocates for science can choose which strategy they prefer. You do not have to be an expert on the content of science to push back against science denial. As Cornelia Betsch put it in an interview, "The problem with topic rebuttal is that you really have to know the science well—and that might be a big ask, because there's a ton of research, and it's sometimes difficult to know everything."[25] By contrast, once you learn the five techniques used by science deniers—and prepare diligently in how to counter them—you can use them as a "universal strategy" for fighting scientific misinformation wherever you find it.[26]

This is great news for those who wish to fight back against science deniers. I felt personally vindicated by this study, because I had already been doing technique rebuttal on my own for more than a year, without quite naming it.[27] But Schmid and Betsch also outline a clear role for scientists to fight back against science denial too. No longer should scientists complain that they are wasting their time when talking to a science denier. How often have you heard a scientist say, "It's not worth talking to these people," or just provide their evidence and then—at the first sign of pushback—walk away? But according to Schmid and Betsch, that is the worst possible thing to do! The evidence in their study shows that such conversations actually can be quite effective, but first we must have them. So what are scientists now to do?[28] Reject Schmid and Betsch's evidence and become science deniers themselves?

Unfortunately, Schmid and Betsch found that while both content and technique rebuttal were useful in mitigating the corrosive effects of scientific misinformation, they were insufficient to counter it completely. Once people had been exposed to scientific misinformation, it had some lingering effect. The best possible thing is for people not to be exposed to any misinformation at all. The worst thing is for misinformation to be shared and not challenged in any way. This leaves the middle ground: if you know that scientific misinformation is circulating, it is better to try to counter it than to do nothing. Here Schmid and Betsch tell a funny story of what to do if you learn of a debate at which scientific misinformation will likely be presented: "not turning up at the discussion at all seems to result in the worst effect. There may be one exception to this: if the advocate's refusal to take part in a debate about scientific facts leads to its cancellation, this outcome should be preferred so as to avoid a negative impact on the audience."[29] Yet even this is sobering. While it appears possible to get people to change their minds after the fact, it is difficult. Even if the method outlined by Schmid and Betsch is helpful, it is not a panacea.[30]

Since we live in a world in which scientific misinformation is ubiquitous, it should come as no surprise that further work remains. For one thing, it is an open question *how* to make use of Schmid and Betsch's results. And there are several potential limitations to their study. For one thing, all they showed is that it was possible to mitigate the effects of scientific misinformation on an audience that had just been exposed to those messages.[31] But presumably some people will have already formed their opinions well in advance, after soaking in misinformation for many decades. One unmeasured factor in their study was whether either content or technique rebuttal had any effect on *hard-core* science deniers, who would have shaped their views based on long-term, repeated exposure to scientific misinformation long before any experiment had begun. It is one thing to intervene immediately and try to keep misinformation from affecting someone's belief. It is another entirely to get someone to change their mind about a core belief that may have hardened over time. Of course, even if further work showed that it was impossible to accomplish this—using any rebuttal tactic whatsoever—it would still be good to know where one's efforts were best placed. Should the target for Schmid and Betsch's strategy consist solely of the still-persuadable *audience* for scientific misinformation? Is it a waste of time to engage in persuasive conversation with a committed science denier?[32] On the basis of their study, we still don't know.

Another possible concern is how any engagement should take place. As noted, Schmid and Betsch's work was all done online. Would their rebuttal strategies still work—or perhaps work even better—if they were employed face-to-face? Here, matters grow more complicated by the need to take into account the social and emotional context that governs personal interactions. As we have seen, we need to factor in the extent to which challenging someone's beliefs in person may involve a degree of identity threat, which might be a factor in any circumstance but especially face-to-face. If we are trying to persuade someone online, it is surely difficult. But in a one-on-one encounter, what additional cognitive, social, and interpersonal factors should be taken into account in shaping an effective rebuttal strategy? As with Kuklinski and Redlawsk, converting people in the lab is one thing, but what to do in real life? As exciting as Schmid and Betsch's work is, we are left adrift on the most exciting practical question.

Once again, we confront the need to consider how incorrect empirical beliefs are actually formed. If people become radicalized not just by misinformation but by surrounding themselves with peers and others who are feeding it to them, might we try to retrieve them through intervention with a different group, whom they could grow to trust to tell them the truth? This intersects precisely with the *absolutely central idea of identity and values to forming beliefs* that we saw in chapter 2. If denialist beliefs are created within a context of identity and values, doesn't it stand to reason that this is how you would change them too?[33]

Dan Kahan tells us that "people acquire their scientific knowledge by consulting with others who share their values and whom they therefore trust and understand."[34] Indeed, it probably works that way for scientists too. So is it that scientists and science deniers just trust different people? Perhaps the goal in talking to a science denier is not merely to give them the facts to make up their mind but to get them to trust scientists again.[35] In his discussion of how best to push back against conspiracy theorists, Mick West talks about the importance of trust and respect in belief formation.

> You can make it clear from the start that you don't really believe their theory, but you can (honestly) say that if there was some compelling evidence, then you would certainly consider it. Give them the opportunity to convert you. This opens the door for them to explain why they believe, and if you genuinely listen to what they say you will gain a very useful perspective, and also increase the odds that they will later also genuinely listen to you. If you respect them, and make an effort to understand their argument, then they will appreciate this, and in turn will respect you more. They will probably have had many situations where their ideas were flatly rejected or laughed at, and so being treated with respect will go a long way toward gaining their trust.[36]

We are here a long way from the information deficit model. Schmid and Betsch provide a solid foundation for the idea that we can change the minds of the *audience* for science denial, just after they have heard scientific misinformation. (Sander van der Linden and others have shown that the same reasoning holds for trying to pre-bunk such beliefs just before people hear them.) But what about those for whom misinformation and disinformation have already shaped their identity? Providing correct information—or challenging misinformation—probably won't work. In some ways Schmid and Betsch's model is still rooted in the idea that the problem with science denial is one of an information deficit. With content rebuttal, we are obviously making up for a lack of information. (Though note that in the process of doing so, we may afford ourselves the opportunity to build trust and respect.) But even in the case of technique rebuttal, we are providing deniers with information about how to reason. We are trying to educate them.

But is any of this really going to change their identity? Maybe, if we do it in the right way. Face-to-face. Over time. With more than one conversation, where we actually listen—which might form into a relationship where our evidence is welcome. That sounds like a lot of hard work, but I think that, especially if we are trying to convert the beliefs of a committed science denier, we must commit to the work of trying to change their identity—probably in person—by trying to build trust through a personal relationship. So Schmid and Betsch are ultimately right about what to do, but the key for actual persuasion might lie elsewhere.

In their previously cited book, *How to Have Impossible Conversations*, Peter Boghossian and James Lindsay explain that as a species we evolved to have conversations—and probably persuade one another—face-to-face. The minute you put information into a text or online, you are taking what might already be a difficult conversation and putting it in "hard mode."[37] And if you want to convince someone, why would you want to do that? Although the authors do not discuss the issue of identity per se, it is implicit that trust and respect are key to getting someone to give up their incorrect beliefs, and that this is best done within the context of a personal encounter.

Engagement, trust, relationships, and values are the keys to real belief change. One should not engage solely in the lab or online. If you meet someone in person, you are building trust. And *then* you can work in your evidence. As Boghossian and Lindsay put it:

> The way to change minds, influence people, build relationships, and maintain friendships is through kindness, compassion, empathy, treating individuals with dignity and respect, and exercising these considerations in psychologically safe environments. It comes naturally to all of us to respond favorably to someone who listens, shows kindness, treats us well, and appears respectful. A sure way to entrench people in their existing beliefs, cause disunity, and sow distrust is through adversarial relationships and threatening environments.[38]

No wonder I wasn't able to convert anyone at the Flat Earth convention. I made the right choice to go there in person, but I needed to listen more. And to make more than one visit. And I should have followed up. No wonder the psychological literature shows such limited success in using facts and reasoning strategies for rebuttal. These are one-off encounters, usually conducted online, in an experimental setting. Yes, those might work, but how much more success might be enjoyed if we used them in person, within the right social context? If we tried to build some trust?

I am convinced that this is the only thing that might work in persuading a genuine science denier. Perhaps nothing will work, in which case our prospects for making any real progress on climate change or vaccines seem pretty grim. But this method has the best shot of working if anything will. It may not be sufficient (as has been shown), but I've come to believe that it is probably necessary.

Changing Minds in the Real World

It is often said that it is stories, not arguments, that actually convince people. In this section, I will share some positive anecdotal accounts of how vaccine and climate deniers changed their beliefs after being exposed to factual information from a trusted source. My goal here is not simply to make the case that science deniers sometimes change their minds—nor to claim that they do so through hearing stories—but to *change the minds of my readers*, who may have doubted that they could play a role in helping science deniers give up their beliefs.

The challenge is that this must happen in the absence of any empirical support, because unfortunately we have outrun the literature. To my knowledge, there is no empirical study that shows the efficacy of face-to-face conversations in convincing hard-core science deniers to give up their beliefs.[39] Of course, we just got done reviewing several empirical studies that suggest that we *can* influence the *audience* for science denial, and that is a very good thing. But where are the studies that show this works on committed science deniers? That test content rebuttal or technique rebuttal outside the lab, where real change often takes place? We encounter science deniers not only online but sometimes in public venues, on the street, or even across the Thanksgiving table. But those studies simply do not exist.[40]

There is, however, a thick anecdotal literature that recounts many cases of staunch, firmly committed science deniers who changed their beliefs. All of these stories are basically the same. They happen within the context of a trusting, personal relationship. As I've said all along, facts and evidence can matter, but they have to be presented by the right person in the right context. That is because, as we now recognize, to change someone's beliefs we do not merely have to fill an information deficit but we have to try to reshape their identity.

Some of the most compelling accounts of first-person science denier conversion have occurred on the topic of anti-vaxx. I haven't said much so far about anti-vaxx. In part, that is because there is an excellent and robust literature on this topic already, and I encourage my readers to consult it.[41] I have even written a bit on this myself.[42] But most people already know the origin of the story.

In 1998, a physician in England named Andrew Wakefield published a small, flawed study in which he alleged that there was a causal link between the MMR vaccine and autism. From the outset, other professionals doubted his work, based on various methodological irregularities and weaknesses in Wakefield's study. As a result, all but one of his coauthors eventually took their name off the study, and the prestigious medical journal that had published it later withdrew it. Indeed, Wakefield's work was so sloppy that he was eventually stripped of his medical license. Later, it was discovered that the problems with the study were not just accidental but the result of outright fraud. By this point, however, it did not matter to thousands of parents of children with autism, who saw Wakefield as a hero who had stood up for them. And as this churned in the popular press, baseless vaccine skepticism grew until anti-vaxx became mainstream.[43]

Despite numerous follow-up studies that found *no link* between vaccines and autism—and thoroughly debunked Wakefield's original finding— thousands of parents began to withhold vaccinations from their children. In 2014, a measles outbreak started in Disneyland and spread across fourteen states. Hundreds of children were infected. A similar outbreak occurred in Brooklyn, New York, and more recently in Clark County, Washington, in 2019.[44] The anti-vaxx movement has continued around the world.

In better news, there are now numerous cases in which former anti-vaxxers have changed their minds. How did this occur? In every case I've read, this happened because someone sat down with the anti-vaxxer, listened to all of their questions, and explained the answers within a context of patience and respect. Some of these stories were already presented in the introduction. Remember those public health officials in Clark County, Washington, who met with parents in small groups, including one physician who spent two hours at a whiteboard explaining cell interaction in a way that was factual but "still very warm"?[45] That worked. Remember Rose

Branigan, who wrote an op-ed in the *Washington Post* about her own anti-vaxx conversion because she found a group of people who were willing to discuss the topic in a kind and rational manner?[46]

In a third account, Arnaud Gagneur, a professor and physician at Quebec's Université de Sherbrooke, engaged in motivational interviews with new mothers in the maternity wing of the hospital. He or a research assistant conducted twenty-minute interviews in which they listened to these parents' concerns and answered their questions. After 3,300 such interviews, they found that mothers were 15 percent more likely to say that they would vaccinate their newborns. Gagneur reports, "They say, it's the first time that I feel respected about my position with vaccination, this is the first time someone has spoken to me like this." One mother is quoted as saying, "It's the first time that I've had a discussion like this, and I feel respected, and I trust you."[47]

With climate denial too one encounters a similar dynamic. Recall Jim Bridenstine—whom we also met in the introduction—who changed his mind on global warming within weeks of starting his new job as chief administrator of NASA? When you begin to have lunch with your adversaries and chat with them in the hallways, miraculous things can happen.[48]

It is worth noting here—although it will surely be dismissed by some as irrelevant and possibly offensive—that in my search for people who have changed their minds on science-denial-related beliefs, I have also run across several instances of those who have given up their ideological beliefs on other, more virulent topics. In one of the most remarkable accounts I have ever read, a young man named Derek Black, a rising star in the white supremacist movement—his father founded the website Stormfront and his godfather is David Duke—went to college and was befriended by a group of Jewish students who invited him to Shabbat dinner every week. Incredibly, this began a chain of events that eventually led him to abandon his views.[49] After beginning a personal relationship with one of the women from the group—who was horrified when she learned of his ideology—she listened and asked questions, then provided material to rebut him point by point. This led to his complete conversion.[50] In a book by Eli Saslow that tells the full story, Black reflects on his conversion:

> People who disagreed with me were critical in this process. . . . Especially those who were my friends regardless, but who let me know when we talked about it that they thought my beliefs were wrong and took the time to provide evidence

and civil arguments. I didn't always agree with their views, but I listened to them and they listened to me.[51]

I am *not*, of course, comparing science deniers to white nationalists. What I am saying is that, if someone can talk another human being out of a decades-long hate-filled ideology like white nationalism/white supremacy simply through listening and friendship, shouldn't it be possible to use the same tactics on a nonhate-related topic like science denial? While reading the account of a different white supremacist conversion, I was intrigued to hear the man's description of what it was like before he changed his beliefs. He said that he was "lost in this ideology" because he felt "marginalized or broken." He explained that his beliefs were deeply rooted in a sense of identity born of isolation, demonization, and hatred that led to grievance against the "other side."[52]

Less dramatic but still important conversions have taken place across the political divide. One former self-described "Trump troll" wrote an account of how he "snapped out of his trance" after being contacted by the liberal comedian Sarah Silverman, whom he had attacked online. He said that she was kind to him and did not argue but instead shared her values. Even though he had been terrible to her, she was respectful to him. And this led to his eventual conversion away from his previous beliefs on gun reform, abortion, and immigration, culminating in his statement, "I now recognize that white privilege exists."[53] Again, I am not trying to draw a link between science deniers and any other ideologies, except to say that to the extent someone's beliefs are based on an information gap, exacerbated by their chosen identity, they may be retrievable through listening, empathy, and respect.

Here it is important to come to grips with the polarizing effect that information sources can have on someone's outlook. If your primary news source about vaccine research is online videos—or your main diet of information about climate change is Fox News[54]—it is easier not only to be misinformed but to demonize or even hate those who are on "the other side." If you have never met a scientist, how will you have the chance to learn that they can be warm and engaging? Even if the primary way to fight science denial is not to provide more information, it is important to be aware that the silo effect can take a toll not only on a person's understanding but also on their tolerance for hearing alternative views that threaten their identity.

Here are a few key things to remember before an encounter with a science denier:

1. Science denial exists on a spectrum If we lump all science deniers together, our task may seem impossible. But there are varying degrees of knowledge and commitment. With anti-vaxxers, for instance, one of the best pieces of advice might be to stop thinking of them all as anti-vaxxers. (And, when speaking to them, we should remember that it is an insult to use the word "anti-vaxxer.") In one Canadian study, it was estimated that hard-core anti-vaxxers make up only 1 to 3 percent of the population, whereas the "vaccine hesitant" make up another 30 percent. Obviously, the latter would be easier to convince.[55]

The same is surely true for climate change. One article from the *Yale Climate Connection* advises that the first step in trying to convert a climate denier should be to differentiate those who are persuadable from those who are not. On the "spectrum of persuadability" they list: informed but idle, uninformed, misinformed, party-line followers, ideologues, and trolls.[56] Obviously, as we already know from Schmid and Betsch's study, the earlier we can get to someone, the better chance we might have. Shall we start with the informed but idle, then work our way through the party-line followers? Since our energy and resources are not limitless, why not focus on where we might do the most good?

But didn't we just get done recounting examples which show that conversion of the hardest of the hard core is possible too? Isn't it worth some effort to try to reach out to the ideologues and trolls? Certainly so, due to their outsize influence.

2. Misinformation and disinformation are amplified on social media Even if there are very few actual hard-core science deniers, they are quite noisy. As one former anti-vaxxer put it:

> Anti-vaxxers have been around for a long time, but social media makes it easier to get into a loop. And once you're there, it's hard to see outside of it. Algorithms just show you more of what you're already looking for. If you start searching anti-vaccination stories, that's what starts popping up on your tagline. You start to think, "Oh, my God, there's all these people and there's so much going on." But if you have a chance to peel back from that, you see that it's actually a very small portion of the population who are really, really loud. The fear makes you angry and it makes you lash out. Once you get into that state, it's easy to stay there.[57]

The spread of science denial may seem bigger than it is, especially if hard-core ideologues are driving the agenda. In *The Conspiracy Theory Handbook*, Stephan Lewandowsky and John Cook offer some sage advice about how best to talk to a conspiracy theorist—by showing empathy and avoiding ridicule—then offer this gem:

> Conspiracy theorists also have an outsized influence despite their small numbers. An analysis of over 2 million comments on the subreddit site r/conspiracy found that while only 5% of posters exhibited conspiratorial thinking, they were responsible for 64% of all comments. The most active author wrote 896,337 words, twice the length of the Lord of the Rings trilogy![58]

Given this, it may be especially worthwhile to occasionally try to engage with the most committed science deniers, for once you have found the (perhaps small) source of disinformation, maybe it can be neutralized. It is also important to remember the shallowness of the disinformation pool. If you engage in some simple interactions with less-committed folks to present information that they aren't hearing from the superspreaders, perhaps you might have a chance to stop more people who start out as questioning but end up as deniers.

3. Be persistent and follow through   Even if you think you have made a successful conversion, you can't just walk away. Again, we can learn from prior research on converting political beliefs. One study found that:

> There was a large bipartisan shift in belief after the fact check, suggesting that both conservatives and liberals can change their minds if they're presented with convincing unbiased information. But there was a catch: After a one-week delay, subjects partially "rebelieved" the false statements and partially forgot that factual information was true. . . . "Even if individuals update their beliefs temporarily, explanations regarding both fact and fiction seemingly have an expiration date."[59]

Updating someone's beliefs based on a fact-check may seem like progress, but full conversion is not just a matter of overcoming an information deficit. The hard work of identity change awaits.

And we know that some people *can* change their minds based solely on evidence. Remember Kuklinski's study, which talked about hitting people "between the eyes"? In 2016, James Cason became the Republican mayor of Coral Gables, Florida . . . and within three days he changed his mind about whether climate change was real. He said, "You know, I'd read some articles here and there, but I didn't realize how impactful it would be of the city that *I'm* now the leader of." Apparently the problem was this: Coral Gables is a wealthy town that has 302 yachts, many of which are anchored at their owners' homes. But there is a bridge between these homes and open water. With climate change, the water level had begun to rise and people were having trouble getting their boats out under the bridge. As Cason observed, "These are $5 million homes with nice boats that suddenly see their property values go down because they can no longer get a boat out. So that will be one of the first indicators [of sea level rise], and a wake-up call for people." Ideology is one thing; not being able to get your boat out of the harbor is apparently an emergency.[60] Later, along with Miami mayor Tomas Regalado, Cason penned an op-ed that appeared in the *Miami Herald* just before one of the Republican presidential debates. It read in part, "For us and most other public officials in South Florida, climate change is not a partisan talking point. It's a looming crisis that we must deal with—and soon."[61]

When self-interest is at stake, people can change their minds fast. Farmers and fishermen now seem to be coming around to accepting the reality of climate change.[62] And even some anti-vaxxers are changing their views due to growing fears about the coronavirus. It's one thing to be anti-vaxx when a disease like measles is relatively rare, and you may not know anyone who has had it; it's another thing to face a life-threatening pandemic with no protection. According to one former anti-vaxxer, "I was just as scared of vaccines as I was of the diseases they protect against. . . . [But] since COVID-19, I've seen firsthand what these diseases can do when they're not being fought with vaccines."[63]

All of this is admittedly anecdotal, but it is still important. The stories of how actual science deniers are converted bear too close a resemblance to one another to be dismissed. In virtually every account I've read, anti-vaxxers, climate change deniers, or other ideologues who have changed their minds have done so on the basis of face-to-face encounters, where evidence was presented by someone with whom they had a trusting relationship. Kindness, empathy, and listening work. These are the keys to helping someone to change their beliefs because they are the route to helping them reshape their identity. And they are also the means by which you might be able to hold the attention of a science denier long enough to work in your data and take care of any information deficit.

As a philosopher who has great respect for science, it pains me not to be able to point to any empirical work that vindicates these speculations. I take heart in the fact that others who have studied this debate, like Michael Shermer and Stephan Lewandowky, have corroborated the advice of Peter Boghossian and James Lindsay that face-to-face respectful conversation is the best way to try to convert someone on *any* subject.

And as far as science deniers are concerned? I talked to Cornelia Betsch on the phone, and she was interested in the possibility of working with me to try to set up a future experiment.

Climate change denial represents the biggest, most important case of science denial in our time. The reason for this is not only that climate deniers are so dug-in and widespread (especially in the United States), but that the costs of inaction are projected to be catastrophic.

The conclusion of a recent IPCC report put out by the United Nations in 2018 was shocking. Not only is the world failing to meet the "can't afford to miss" target of 2 degrees Celsius temperature increase overall—which would require cutting global emissions beyond their current level—but in 2017 the global emission of carbon dioxide *increased* to the highest level on record.[1] Then, in 2018, it went up again.[2] The full data for 2019 are not yet available, but based on preliminary analysis they are expected to hit another all-time high.[3] "We are in deep trouble with climate change," concluded UN Secretary-General António Guterres at the opening of the twenty-fourth annual United Nations climate change conference in Poland in 2018. China saw a nearly 5 percent emissions growth in 2018, joined by 6 percent in India. As the second-largest producer of greenhouse gases (exacerbated by Trump's coal-friendly agenda), it is perhaps not surprising that the US saw a 2.5 percent increase in its carbon dioxide emissions in 2018. To make matters worse, scientists now project that even the 2 degree Celsius goal (as set out in the earlier Paris Agreement) is not ambitious enough. In order to prevent the worst effects of global warming, the world should instead be aiming at no more than a 1.5 degree C increase.[4]

Perhaps the most alarming conclusion, though, was that if we continue at our current rate, we will reach the 1.5 Celsius target by 2040. (We are already two-thirds of the way there; the global temperature has gone up 1 degree Celsius since the industrial revolution in the 1850s.)[5] Beyond that, if we do nothing, the global temperature is expected to rise by anywhere from 3 to 5 degrees Celsius (5.4 to 9 degrees Fahrenheit) by the end of the century, which would be devastating.[6] By that time, the worldwide economic costs of climate change will amount to more than $54 *trillion*.[7] This includes infrastructure damage, lost wages, property loss, and other economic costs. But the human and social costs would be even more tragic and incalculable. Environmental effects like extreme heat, increased wildfires, flooding, hurricanes, water shortages, and crop loss are expected to result in millions of heat-related deaths, climate refugees, and social collapse on a scale the world has never witnessed.[8]

The slim piece of good news is that *if* we cut global carbon dioxide emissions by half by the year 2030, we could still hit the 1.5 degree target.[9] (But in order to maintain this, we would then have to cut them to zero by 2050.)[10] Yet, as the IPCC report bluntly puts it, "there is no documented historic precedent" for the sweeping change necessary to energy, transportation, and other systems required to reach 1.5 degrees Celsius.[11] Could technology help? Absolutely. As a matter of fact, according to Princeton University scientists Robert Socolow and Stephen Pacala, "humanity already possesses the fundamental scientific, technical and industrial know-how to solve the carbon and climate problem for the next half-century."[12] Could economic incentives make a difference? Yes. A worldwide carbon tax might provide the catalyst to modify our consumption habits so that we would make better environmental choices. Most important, we would have to stop using coal.[13] Although this would be painful, there is every incentive to embrace these costs *now*, while they are lower, rather than defer them to the future. As UN Secretary-General Guterres stated, "It is hard to overstate the urgency of our situation. . . . Even as we witness devastating climate impacts causing havoc around the world, we are still not doing enough, nor moving fast enough, to prevent irreversible and catastrophic climate disruption."[14]

Yet how likely is it that any of these things will happen without the political will to bring them about? We already live in a world in which even those countries run by politicians who acknowledge climate change are missing their targets. In 2018, French President Emmanuel Macron faced massive rioting in his country for imposing a modest fuel tax, which he later rescinded while saying, "No tax is worth jeopardizing the unity of the nation."[15] Previously he had said, "One cannot be on Monday for the environment and on Tuesday against the increase of fuel prices."[16] Nonetheless, he blinked.

In the US, we have a political leader who is incapable of blinking, because he has already shut his eyes to the reality of climate change. As has been widely reported, President Trump pulled the US out of the Paris Agreement at the first possible opportunity (which was November 2020).[17] In the meantime, he has done everything he can to talk down the issue of climate change. He reinstated subsidies for coal production.[18] He rolled back Obama-era emissions standards on new cars.[19] In the face of the 2018 wildfires in California, he dismissed the idea that they could have anything to do with climate change, and recommended that firefighters spend more time "raking the forest floor."[20] In an October 2018 interview where Trump was confronted by a reporter who asked whether he still believed that climate change was a hoax, they had the following relatively unbelievable on-camera exchange:

STAHL: Do you still think that climate change is a hoax?

TRUMP: I think something's happening. Something's changing and it'll change back again. I don't think it's a hoax, I think there's probably a difference. But I don't know that it's manmade. I will say this. I don't want to give trillions and trillions of dollars. I don't want to lose millions and millions of jobs. I don't want to be put at a disadvantage.

STAHL: I wish you could go to Greenland, watch these huge chunks of ice just falling into the ocean, raising the sea levels.

TRUMP: And you don't know whether or not that would have happened with or without man. You don't know.

STAHL: Well, your scientists, your scientists—

TRUMP: No, we have—

STAHL: at NOAA and NASA—

TRUMP: We have scientists that disagree with that.

STAHL: You know, I—I was thinking what if he said, "No, I've seen the hurricane situations, I've changed my mind. There really is climate change." And I thought, "Wow, what an impact."

TRUMP: Well—I'm not denying.

STAHL: What an impact that would make.

TRUMP: I'm not denying climate change. But it could very well go back. You know, we're talking about over a millions—

STAHL: But that's denying it.

TRUMP: of years. They say that we had hurricanes that were far worse than what we just had with Michael.

STAHL: Who says that? "They say"?

TRUMP: People say. People say that in the—

STAHL: Yeah, but what about the scientists who say it's worse than ever?

TRUMP: You'd have to show me the scientists because they have a very big political agenda, Lesley.[21]

Amid other US politicians making wild speculations about "global cooling" or other dismissive comments about "alarmism" among climate scientists,[22] one must also confront the rank cowardice of Trump's decision to release a government-mandated congressional report on climate change on Black Friday—the day after Thanksgiving—in 2018, so that reporters might miss the government scientists' own unvarnished conclusion that, if something isn't done soon, the United States will face an economic hit of 10 percent of its annual gross domestic product (GDP) by the end of the century.[23]

The Flat Earthers may have seemed harmless, but this kind of science denial could kill us.[24] Fortunately, with each passing year, there seem to be fewer holdouts on the issue of global warming. According to a 2018 Monmouth University poll, 78 percent of Americans now believe in the reality of climate change and agree that it is a "very serious problem." Yet this figure includes only 25 percent of Republicans. And overall public opinion remains split on the causes of climate change, with only 29 percent of respondents accurately reflecting the 97 percent consensus view of the world's scientists that human activity is nearly completely responsible for the increase in global temperature.[25] Surely not all of these people are science deniers. Some may just be ignorant of the facts, which have been obscured by a campaign of disinformation that has come down to us through corporate lobbying and political self-interest. But this is why it is so important for those of us who understand the importance and consequences of climate change, and trust the scientists who are studying it, to continue to talk about it.

The Origins and Causes of Climate Denial

There is a massive gap between public understanding and the reality of climate change. Among our elected leaders, this gap is perhaps most challenging, given the imbalance in power between Democrats and Republicans during the Trump presidency, and the widespread foothold of climate denial in the Republican Party.[26] If we are going to do anything about climate change, the US must take a leadership role. But without the political will to see this happen among our citizens, what chance is there that our elected leaders will follow suit (or that we will get new leaders who will)?

Now to the unavoidable question: if the science is so clear about climate change, why are there so many people who deny it? Here I trust it is not necessary for me to present the evidence to show that climate change is true. There are ample resources elsewhere.[27] Just as I understood that it was not my job to prove to the reader that the Earth is actually round in chapter 1, it should not be my goal here to present the scientific evidence for climate change. What does need looking into, though, are the motivations and strategies pursued by climate change deniers. Why do I call them "deniers" rather than "skeptics"? Because the evidence is so clear, and there is such great consensus among scientists, that it isn't even a close call.

In a 2004 study, Naomi Oreskes examined all 928 papers on the topic of "global climate change" published between 1993 and 2003, and found that exactly *none* of them disagreed with the consensus scientific position on global warming.[28] In a 2012 follow-up, James L. Powell found that of 13,950 peer-reviewed articles on climate change from 1991 to 2012, only twenty-four of them (0.17 percent) rejected the idea of global warming.[29] In a 2014 update of 2,258 more articles, Powell found only one additional paper that challenged the scientific consensus on climate change.[30]

It is often said these days that 97 percent of scientists agree that climate change is happening and that human activity is primarily responsible. One source for this claim is a 2009 survey by Peter Doran and Maggie Zimmerman, which found that among climate science specialists, 96.2 percent agreed that global temperatures have risen since 1800, and 97.4 percent agreed that human activity was a significant contributing factor in changing global temperatures.[31] This finding was corroborated in a 2013 paper by John Cook et al., which examined the abstracts of 11,944 papers on "climate change" or "global warming" published between 1991 and 2011, and found that—of those that expressed an opinion—97.1 percent agreed with the consensus scientific position on anthropogenic global warming.[32] Then, in a stunning follow-up, in 2015 Rasmus Benestad and colleagues examined the work of the proverbial 3 percent of scientists who rejected global warming—which consisted of thirty-eight papers that had appeared in peer-reviewed journals in the previous decade—and found that *all of them* were methodologically flawed![33]

The upshot is that there is virtually *no* scientific debate on the reality of climate change. Or, rather, the amount of dissent is exactly what one would expect on any scientific question where there is a mountain of evidence but no possibility of proof: miniscule.[34] There is, however, a huge disparity between what scientists believe and what the public *thinks* scientists believe. That is, the public is misinformed not just about whether climate change is happening (and who is responsible) but about whether scientists *agree* that climate change is real (and that humans are overwhelmingly responsible for it). This is the result of a campaign of disinformation that has been manufactured by both corporate and political interests and caused the public to be misled.

As Dana Nuccitelli (one of the authors of the Benestad study) put it:

> When queried about the most recent IPCC report, Republican lawmakers delivered a consistent, false message—that climate scientists are still debating whether humans are responsible. The previous IPCC report was quite clear on this, attributing 100% of the global warming since 1950 to human activities. As NASA atmospheric scientist Kate Marvel put it, "We are more sure that greenhouse gas is causing climate change than we are that smoking causes cancer."[35]

"Yet," Nuccitelli goes on to report, "as surveys by Yale and George Mason universities have found, only about 15% of Americans are aware that the expert climate consensus exceeds 90%."

How did this happen? The political story here is preceded by an economic one, due to corporate interests that are not dissimilar to what happened with smoking and lung cancer in the 1950s. Remember that meeting of tobacco company executives in the 1950s, when they came up with a plan for fighting the science on smoking and cancer? An almost identical meeting happened forty years later when the American Petroleum Institute (with members such as ExxonMobil, BP, Chevron, and Shell Oil) convened in 1998 to create a "Global Climate Science Communications Plan" to fight the science on global warming. With the 1997 Kyoto Protocol meeting having already gotten many of the world's nations to commit to lowering carbon emissions, it was time to act. This resulted in a strategy to create confusion over the science behind climate change, following the same blueprint that the tobacco executives had used years before. Recall the infamous tobacco memo that said, "Doubt is our product." This time the API action plan got leaked, but it didn't take decades to surface.[36] It said, "Victory will be achieved when average citizens 'understand' (recognize) uncertainties in climate science" and "those promoting the Kyoto treaty on the basis of extant science appear to be out of touch with reality."[37] But this time it really didn't matter. The battle lines were clear and a leaked memo didn't have the same punch. Years later, it was learned that ExxonMobil knew about the reality of climate change as early as 1977.[38] Indeed, in the very definition of hypocrisy, ExxonMobil was making plans to explore new oil fields in the Arctic once the polar ice cap had melted, even while it was ramping up efforts to foment climate change denial.[39]

It may be hard to remember, but it wasn't always like this. When global warming first came to public attention in the late 1980s, President George H.W. Bush promised to fight the "greenhouse effect" with the "White House effect." One result was the creation of the Intergovernmental Panel on Climate Change, which has done so much to raise public awareness about global warming.[40] Even as late as 2008, there was still some semblance of bipartisanship: witness a public service announcement on television, where Republican Newt Gingrich and Democrat Nancy Pelosi sat on a couch and promised a unified approach to fight global warming.[41] Of course, by then Al Gore was already back in the spotlight with his slide shows, culminating in his 2006 book and film, *An Inconvenient Truth*. The issue of climate change was already on its way to being politicized, even though it had not yet reached the level of partisan fervor we find today. First, the politicians would have to completely capitulate to corporate interests, who had a vested stake in seeing how this "debate" turned out.[42]

In her 2016 book, *Dark Money*, Jane Mayer makes the case that climate denial was initiated by those with fossil fuel investments, such as the Koch brothers and ExxonMobil.[43] Indeed, the money coming from Charles and David Koch alone was staggering: "From 2005 to 2008, a single source, the Kochs, poured almost $25 million into dozens of different organizations fighting climate reform."[44] This meant that they outspent ExxonMobil by a factor of three. Elsewhere Mayer writes, "If there is any lingering uncertainty that the Koch brothers are the primary sponsors of climate-change doubt in the United States, it ought to be put to rest by the publication of *Kochland*."[45] Others have agreed that "few humans hold more responsibility for the unfolding climate crisis than David Koch."[46]

Of course, there was money from fossil fuel companies as well. In a 2019 *Forbes* article it was revealed that:

> every year, the world's five largest publicly owned oil and gas companies spend approximately $200 million on lobbying designed to control, delay or block binding climate-motivated policy. . . . BP has the highest annual expenditure on climate lobbying at $53 million, followed by Shell with $49 and ExxonMobil with $41 million. Chevron and Total each spend around $29 million every year.[47]

All told, *Smithsonian Magazine* estimates that "nearly a billion dollars a year is flowing into the organized climate change counter-movement."[48]

What does all of this money buy? Think tanks.[49] Conferences.[50] Lobbyists. Research by industry-friendly scientists. Media coverage. In a word: doubt. In *Dark Money* Mayer reports that Harvard political scientist Theda Skocpol marks 2007 as a turning point in the fight over climate change. This was just after Al Gore was awarded the Nobel Peace Prize and *An Inconvenient Truth* had come out. Polls showed that the public was beginning to worry more about global warming. At this point, the forces of climate denial began to fight back with increased vigor. Radio, TV, books, and testimony before Congress all contributed to a public relations push for more skepticism about climate change. Skocpol estimates that during this period climate denial was disseminated to between 30 and 40 percent of the US population in their daily media.[51] And the result was predictable: "before long, public opinion polls showed that concern about climate change among all but hard-core liberals had collapsed."[52]

The effect on US politicians was predictable as well. With shifting public opinion, not to mention millions of dollars in political contributions from fossil fuel interests over the years, "the Republican party, particularly in the U.S. Congress, soon swung sharply to the right on climate issues. Partisan differences remained small among the general public but grew into a gaping chasm among elected officials."[53] In an interview about his book *Kochland*, Christopher Leonard said:

> The Koch network played a vital and unrivaled role in burning down the moderate wing of the Republican Party that acknowledged the reality of climate change. And that forever changed the political discourse, to where now, any credible Republican politician who wants to raise enough money for re-election cannot even acknowledge the basic facts of science.[54]

Today, we find ourselves with the legacy of all this denial and disinformation. Even as public opinion on the reality of climate change has started to rise again, politicians sit pat.[55] Why? Because the newest form of denial isn't over the *existence* of climate change but whether we know enough that we can (or should) try to do anything about it.[56] And this is more partisan than ever.

In the latest Pew poll, researchers found that for the first time a majority of Americans (52 percent) said that taking some action on climate change should be a top priority for the president and Congress. But there is a deep partisan divide. In the past four years, Democrats' support for more policy action on climate change has grown from 46 percent to 78 percent, but Republicans have remained essentially unmoved from 19 percent to 21 percent.[57] Rather than a question of knowledge, it seems that belief in climate change has become a matter of identity.[58] Just like Flat Earth, it is not even about the evidence anymore—it's about which team you're on.

Now, how can I say all this? Isn't there any grounds for debate on climate change as a scientific question? What happened to good old-fashioned skepticism? Is the evidence for climate change really just as clear as that for the shape of the Earth? If so, why can't they *prove* it? Ah, but here we are back in a familiar place, and you've seen this script before. As we know, everything in science is open to dissent. There is always an alternative hypothesis that *could* be true, but this doesn't undermine warrant. The amount of evidence for anthropogenic global warming is *gigantic*. Remember that Reuters story about how the evidence for climate change has now reached the 99.9999 percent confidence level?[59] In the face of overwhelming evidence, it is irrational to disbelieve something just because the alternative *might* be true. Are there striped unicorns at the South Pole? How do we know if we've never been there? But this is Flat Earth all over again. The rejection of such a huge amount of scientific evidence, and scientific consensus, is not skepticism, it's denial.

So why have we let fossil fuel interests and conservative politicians so effectively exploit the issue of doubt, as if they had discovered some enormous flaw in scientific reasoning? It's time to push back. The antidote to climate denialists is to expose the full nature of their financial and ideological corruption, and the similarity of their argumentative strategy to that used in other trumped-up denialist movements about evolution, vaccines, and the shape of the Earth. But here an interesting question arises. We understand that climate denial has a lot of similarities to the campaign to deny the link between cigarette smoking and cancer, but what does it have in common with something like Flat Earth? I can't for the life of me figure out who is benefiting from Flat Earth. Yes, some people who have jumped on the bandwagon may make a dollar or two, or even a living, with their T-shirts, websites, hats, and books. But was Flat Earth cynically created for that purpose? I doubt it.

Sometimes science denial is like that. In some instances it can arise from obvious self-interest, but in others it can arise seemingly out of nowhere. With climate denial, there is little doubt how it was created. The work of Oreskes, Conway, Leonard, Mayer, Mooney, Hoggan, Coll, and others has made clear how deeply cynical and corrupt the origins of climate denial have been, and how it has been whipped over the years by those who either profit from it or are beholden to those who do. Yes, some people actually believe it. They are the pawns in this story. But does this mean that the folks who created it do not also perhaps believe it too?

In an interview about *Kochland*, Christopher Leonard reflects on the question of whether David and Charles Koch actually believed in the truth of what they were doing, even while they were funding and spreading misinformation.

> I'm not trying to sound cheesy here; this is the honest truth. . . . As an outside journalist, I cannot sit here and give you the satisfactory answer. Do they really believe that climate change is . . . phony or overstated? Do they really believe that market forces are going to come in and solve this problem? . . . Charles Koch would not talk to me on the record or answer my questions on this topic; the company would not make him available. But I will tell you, I interviewed senior, senior Koch Industries people who'd been there for decades, who believed in their heart that climate change is a hoax. So I don't know how much of this is a belief system that's reinforced when you live in the oil industry world, and how much of it is intentionally avoiding the scientific evidence. I just don't know.[60]

Does this mean that climate denial does not follow the same script as Flat Earth? That the playbook is different, depending on whether it was cynically created or organic, whether you are trying to mislead other people, or actually believe in it yourself? Absolutely not. Climate denial still follows the five tropes playbook, just as surely as Flat Earth did. Even though the five tropes of science denial reasoning were probably not consciously designed to get people to believe in Flat Earth, it is still the backbone of their reasoning. In the same way, even though climate denial *was* diabolically created by those with corporate and ideological interests, it follows the same playbook. This is a preexisting scheme left over from the tobacco strategy of the 1950s, and it conveniently fits virtually all forms of science denial. So we should not be surprised that climate denial fits this classic pattern:

Cherry-Picking

Yes, absolutely. We already saw this with Ted Cruz and others, who deliberately chose 1998 as the base year for a bogus claim that global temperatures did not increase over an eighteen-year period. Notably, even when this was debunked, they continued to make the claim.[61]

Conspiracy Theories

Of course. President Trump has made numerous claims over the years about how climate change is a hoax perpetrated by the Chinese to hurt American manufacturing, that the scientists are politically biased, etc.[62] During the 2009 Climategate debacle, deniers seized on some inappropriate emails sent by scientists at the University of East Anglia, and tried to use them to show that there was a worldwide conspiracy of climate scientists.[63]

Reliance on Fake Experts

Here the issue is a tad more subtle. Some of the work cited by climate skeptics has been done by actual scientists (some with credentials in climate science), but their work has been cherry-picked with great preference for preexisting views that are hostile to climate change. In some cases, climate deniers have been reduced to relying on folks with no credible credentials in climate science; in others, they've focused on those who may agree with the scientific consensus that climate change is real and that humans are causing it, but doubt that global warming is a very big deal. These scientists are then feted at anti–climate change conventions, where they are treated like rock stars. If you'd like to go, these conventions seem to be an annual offering, with one coming up in Las Vegas in April 2021.[64]

Illogical Reasoning

Examples of this are numerous, and we have already encountered one. Remember the strawman claim from chapter 2, where the climate denier said that human activity was not the *only* cause of greenhouse gas emissions? That may be, but it is the main driver.[65]

Insistence That Science Must Be Perfect

We just examined this. Climate deniers routinely say that projections about global warming are just models and that if there is any uncertainty or error, we should wait for more evidence. Of course, this is a ridiculous standard and they know it. This is denialism 101. Exploit any doubt, no matter how small, as a legitimate reason for disbelief.[66] Then capitalize on the delay while you profit.

Climate skepticism is thus not actually skepticism at all. In the face of an avalanche of evidence, continuing to shill for a contrarian point of view just because you hope it will be right is outright denialism. It may seem more reasonable than Flat Earth, but it is not. It relies on manufactured doubt that is whipped through the processes of disinformation and distortion that were created to keep us from drawing a reasonable conclusion despite scientific consensus. This is the identical strategy used by Flat Earthers. As we've seen many times now, all science denial is basically the same.

The Maldives: Ground Zero for Climate Change

In March 2019, my wife, Josephine, and I traveled to the other side of the planet to the tiny island nation of the Maldives. The Maldives is the lowest and flattest country on Earth, plopped right in the middle of the Indian Ocean, about 600 miles off the coast of Sri Lanka. Its average elevation is four feet above sea level; its high point stands eight feet above sea level, which is the lowest high point of any nation in the world. Home to about 500,000 people, spread out over 1,200 islands (200 of which are inhabited), the Maldives is perhaps the most vulnerable nation in the world to the consequences of climate change. As such, it plays a prominent role in bringing the threat of global warming to worldwide attention.

The primary risk is flooding. If current trends continue, the landmass of the Maldives will literally disappear, and its inhabitants will have to be evacuated and relocated to another home, by 2050. As Tony deBrum, former foreign minister of the Marshall Islands, put it, "Any increase beyond 2 degrees is a death warrant for our countries."[67] Even before the islands disappear, however, they may become uninhabitable. The only source of fresh water in the Maldives is rainfall, which resides in ground aquifers. With storm overwash from the sea, these become contaminated with salt, and eventually it will reach the point where rainfall later and later each year is not enough to flush it out. At that point, a decision will have to be made.

Maldivians are ingenious people, and one of their first moves years ago was to bring their plight to public attention. Their fourth president (and the first democratically elected one), Mohamed Nasheed, held the world's first underwater cabinet meeting (complete with scuba gear and a camera crew) to capture the world's attention. He also starred in a documentary film called *The Island President* that recounted the nation's efforts to get the world to agree to the 2 degrees Celsius goal at the Paris climate meetings. Along the way, he committed to making the Maldives carbon-neutral as an example to the wealthier (and more polluting) countries. Even though Nasheed was ousted in a coup in 2012, one of his main strategies to address climate change has survived, as the Maldives has embarked on a plan to save for its future. Skeptical that the world's other nations will do enough to rescue them, the Maldivian government is putting aside a significant portion of its annual $2 billion tourist revenue into a sovereign wealth fund to buy a new home somewhere in the world, against the day when the entire nation will have to relocate.[68]

We left Boston at eleven p.m. for the overnight flight to Dubai, where we had a seven-hour layover before the next flight to Malé, capital of the Maldives. Malé is one of the densest cities in the world, with nearly a quarter million people crammed into just over two square miles. On the flight in, you can see the entire island of Malé, squat and teeming, in the middle of the bluest water you'll ever see in your life. During our time in the Maldives, we witnessed this time and again, but with opposite density. While Malé accounts for almost 40 percent of the nation's population, the rest is scattered over thousands of tiny islands that extend over 35,000 square miles, of which only 115 square miles is land. The Maldives is truly an *island* nation, with the only practical means of transportation by boat or plane.

Each inhabited island is something of a world unto itself, yet all must conform to strict laws concerning manner of dress, alcohol, and public displays of affection, in this 100 percent Muslim country. In addition to the 200 inhabited islands, there are 132 "tourist islands" (which are considered uninhabited because no Maldivians live there). These are run by the hotels and tend to have looser rules—about bikinis, pork, and public drinking—but there is still a government official for each island, who ensures conformity with local laws.[69] This arrangement means that there is more or less strict segregation between locals and tourists. There are virtually no hotels on islands that are inhabited by Maldivians, and no Maldivians (other than hotel staff) who live on the tourist islands. In fact, it is common for each of the large hotels to have its own private island. One hotel, one island. So there are 1,200 islands, 200 inhabited by Maldivians and 132 run by hotels.

One of the most interesting sights in Malé, which you can see from the ground at the international airport (which is on its own island, Hulhule), is a twin island being dredged up from the sea right next to Malé called Hulhumalé. Hulhumalé—or "New Malé"—is an artificial island being built in response to overcrowding on Malé, and the growing threat of climate change. Imagine living in Malé, knowing that someday a wave—or simple flooding—could overtake one of the densest cities in the world.[70] Where would you go? Hulhumalé is already inhabited and has a developing economy of its own, but it is being built primarily as a "bugout" location, against the day when Malé is threatened. The average elevation on Malé is three feet. Hulhumalé is being built to a height of six and a half feet. Amid worries of the effects of climate change in American cities like New York, imagine if the response was to build an entirely new (slightly taller) island called "New Manhattan" right next to it.

The many cranes atop the buildings on Hulhumalé suggest a feverish pace of construction. The huge sand piles at water's edge made me think of the history of the Back Bay in Boston (near where I live) and other land-reclamation efforts. But will this one last? Hulhumalé is a stopgap, built for the temporary relief of those who will be displaced from lower-lying islands due to climate change in the coming years. It is an insurance policy. But will it be enough? The funneling of so many tourist dollars to the sovereign wealth fund suggests not, as someday even Hulhumalé will likely be overtaken by water. At that point, the residents will have to escape to Sri Lanka, India, Australia, China, or the United States as climate refugees, unless they are able to find a nation that will sell them enough land to reconstitute the Maldives elsewhere.

Our second flight was to the island of Kooddoo on a tiny propeller plane. From the air we could see islands everywhere, connected in rings known as atolls, which were many miles apart. Thick white clouds filled the sky as we looked down at the bright blue water. Once we arrived in Kooddoo, we had a short wait for the boat that would take us to our destination island of Hadahaa. We arrived in Hadahaa approximately thirty-six hours after leaving Boston, and even though I was sleep-deprived, I was more excited for this trip than any other I'd taken in my life. You know how after traveling for many years, every place begins to look the same? Not this.

The boat pulled up to the dock in Hadahaa, and we couldn't stop smiling. There were no roads. The entire island is so small that you can walk around it in twenty minutes. There are no other islands nearby, though you can see some in the same atoll on the far horizon. Hadahaa sits thirty-four miles from the equator, near the southern tip of the Maldives. We had wanted to go as far out as possible, because the outermost islands are the most vulnerable to climate change, and had arranged to stay at a resort that had a resident marine biologist, with whom I planned to study the local effects of climate change.

Now, I will not lie to you: Hadahaa is a wonderful resort. It is one of those islands that are owned and operated by a hotel, where a visitor is treated to first-class service and some of the most beautiful sights you will ever see in your life. We were among those couples paying a shocking amount toward taxes for the sovereign wealth fund—and we also offset the entirety of our trip with carbon credits when we got back—so all of the money went to a good environmental cause. But I was there to work. My goal was to understand not just the physical but also the cultural effects of climate change, so I needed to be there firsthand. In addition to working with a marine biologist, I also chatted with native Maldivians who worked at the hotel, spoke with fishermen, and took a snorkeling excursion to see coral death in the middle of the Indian Ocean. But we didn't have to go that far to see the effects of climate change. There was dredging due to erosion on one of the beaches on Hadahaa, where the hotel was trying to contend with the fact that the beach was slowly shrinking. There were life vests in the closet of our room, and there were sandbags aplenty lying just under the sand on the beach. During a specially arranged back-of-house tour, we also got to see how staff lived on the island.

Hadahaa boasts two hundred staff for one hundred guests. The staff live in dorms, three to a room, in the center of the island, with free room and board, health care, thirty days of vacation, and shared tips. They are mostly young people, there for a year or two, some of whom send money back home. By law, half of the staff are native Maldivians, but the others come from nearby countries. It is immediately clear that—despite a relatively high annual income in the Maldives compared to neighboring countries—there is great wealth inequality here. Hotel jobs are coveted. I also struggled with the troubling fact that these islands—which are at greatest threat of climate change of any place in the world—are forced to run a vast amount of air-conditioning equipment and use many other resources for their guests, which surely contributes to global warming. But that is just reality in the Maldives. There are really only two industries: tourism and fishing. And 90 percent of the tax revenue in the Maldives comes from tourism, without which they could not afford to contemplate survival of the effects of global warming. It is ironic that Maldivians have to use so many resources to generate the income to save themselves from the effects of their guests' consumption of those very same resources, but is this really the Maldivians' fault? Surely not. If the great industrialized countries in the world have been exploiting carbon resources for hundreds of years, which led to the accumulation of great wealth, why should we expect the Maldives and other underdeveloped countries to be the ones to carry a special burden and save us from climate change? And yet they try. With the aforementioned goal of carbon neutrality, on Hadahaa they were constantly working toward less waste. During our tour, we saw the installation of a solar farm and water-recycling efforts. It is hard not to feel some guilt at being in the Maldives, enjoying a luxury resort, especially when the point of the journey was to learn about climate change. But as long as we were there, we enjoyed the beach, tipped generously, and got ready for the research component of our trip.

We met our marine biologist, Alex Mead, at the dive shop, where we had to be outfitted with wet suits, life jackets, and snorkeling equipment before the trip. Alex was a certified dive instructor with many years' experience and also a scientist with training at the University of Plymouth in England. I was immediately struck by his more than passing resemblance to Henry Golding, the lead actor in the popular film *Crazy Rich Asians*, which had just come out the previous summer. Tall and athletic, he even spoke with a British accent as he explained his background.

Our first order of business was to attend a short presentation on the role of coral reefs in island formation. The Maldives started with volcanic rocks that sunk to form a barrier reef. Eventually, this sunk too and left behind a ring of islands that formed the atolls. These were formed by live coral (which are animals, not plants) that built up to the waterline. They can't go any higher than this, but can go right up to it, so the higher the water, the higher the coral. Fish then build on the coral with sand from their excretions and use their teeth to gnaw away dead material. Sand slowly builds up and offers some protection to what lies behind it. Over time, this allows flora and fauna to grow and form an island. All of this takes about 10,000 years, but it could be gone in about fifty years (or sooner) due to global warming.

There are two main threats from climate change.[71] One is that warmer water eventually leads to breakup of the coral reefs. It literally kills the coral (which turns it white), and then there is no protection for the island, as the beaches erode. Eventually, the island itself will break up. This is completely due to water temperature and has nothing to do with rising sea levels. But the second danger comes from overwash—which *is* due to rising sea levels—when the islands are overtaken by water. In extreme cases this can come in the form of floods or a tsunami, which become more common as global temperature increases. As more storms occur, there is more possibility of overwash: a wave rolls over the low-lying ground on the island and wipes out everything in its path. So it's not just flooding but extreme storms that can cause erosion. And as these become more common, the islands become uninhabitable.[72]

Now it was time to go to the boat and see all of this firsthand.

Our crew consisted of three skinny young Maldivian men, who were barefoot and at first said they didn't speak any English. They appeared to be in their late teens. They ran the boat while Alex led the tour. Our first stop was the island of Nilandhoo. This was home to about a thousand people, all native Maldivians. What a shocking contrast with the resort island! Although the scenery was gorgeous, the quality of life for the locals must have been quite difficult. There were abandoned buildings with graffiti and one dirt road that traveled from one end of the island to the other. Standing in the middle of the street at high noon, looking first one way then the other toward the sea, it was easy to see why climate change was such a threat. Alex said that one high wave could easily wash from one side of the island to the other in a flash.

There was zero traffic on the street, aside from a couple of bikes. There were no cars. The men were shirtless, but the women wore long, dark robes and hijabs on this nontourist island. We stuck our heads in the local clinic, which looked clean but had old-fashioned glass IV bottles. We didn't go inside. We attracted quite a lot of attention the entire time we were on the island. Although we were modestly dressed, visitors must have been a rare sight on Nilandhoo, and one man on a bike gave us a hard stare the whole time our crew navigated the boat out of the harbor.

Our next step was the highlight (and also the scariest part) of the trip for me: snorkeling off the Doragalla Thila. This is where Alex said he'd take us to see the dead coral. At first I couldn't understand how we were going to see any coral in the middle of the ocean. Coral was near the shore, right? But I suppose the seafloor varies in height—which is why there were islands in the middle of it, after all—so I put on my snorkel gear and enjoyed the ride. The crew seemed to enjoy speeding away from Nilandhoo as fast as they could in open water, but all of a sudden they slowed to a stop. I couldn't see anything. We were very far from shore, and the water looked a uniform blue when Alex announced, "This is the spot."

Alex of course was the first one into the water. My wife was next. Now, I am not a strong swimmer, and in fact I only learned to swim when I was in my early thirties. At that time, I was a new father and wanted to learn to swim because if my kid ever fell into the water, one way or another I was going in after her. My swim instructor at the small liberal arts college where I taught understood my motivation, so he put me in a class with the fraidy-cat seniors, who needed to pass the swim test to graduate. After three months of practicing in the shallow end—or staying near the side of the pool when I had to go in the deep end—I learned to swim. But then came my big moment. I'd vowed to become the first faculty member in school history to take the swim test with the seniors. The test involved jumping into the deep end, then swimming four laps, treading water for five minutes, then swimming four more laps. I remember hesitating at the edge of the pool. The instructor said, "Look, Lee, you already know how to do this, but remember why. Your daughter is in the water. You've got no time to think. Now jump." I said, "You bastard," and jumped. Then I passed the test.

But that was twenty years ago, and now I was in open water in the Indian Ocean. There wasn't any land nearby. For a moment I hesitated at the side of the boat. Suddenly, watching my wife and Alex drift farther out into the current, I felt overcome by a sense of urgency. If I waited much longer, I'd have to swim farther out to reach them or risk not going in at all, in which case the whole purpose of my trip would be blown. Whether it was that—or the fact that my wife of thirty-two years was already in the water with an impossibly handsome nautical version of James Bond—I jumped.

Suddenly I was in trouble. My life jacket felt like it was tipping me forward. Josephine called out, "Come on, Lee. You can make it," and waved. I started swimming. Breathing through the snorkeling tube felt odd, but at least I was moving. I kept looking up to see if I was making progress. Finally I reached them and put out a hand. Then I heard an engine start . . . and the boat left!

I was in the middle of the ocean—with a certified dive instructor and my wife—but nothing else. I couldn't even go back to the boat if I got scared.

"Alex, what happened?"

"What . . . what do you mean?"

"Where's the boat going?"

"Oh, they'll be back."

"When?"

"Are you all right?"

"Yes, but I'd like to know where the boat went? Do you have a set time for them to come back? Did something go wrong? Did you make arrangements?"

Alex hesitated for a moment and smiled. "No, they always do that. Not to worry. They have to leave or risk getting the bottom of the boat hung up in the shallows with the coral. I can call them back anytime. It's all right."

I calmed down slightly. My wife was already snorkeling, and Alex said, "Come on, we're not far from the reef. I've got lots to show you."

I nodded and put my face in the water again. And I was mesmerized. As we approached the reef, I saw that there were fish galore! Clownfish. Parrotfish. The most remarkable panoply of fish I'd ever seen outside a Disney movie.

Then I saw Josephine gesture. It was a shark. Alex had told us about the possibility of blacktip reef sharks. He'd said they weren't dangerous, even if they looked it. More dangerous were the black eels, which were rare.

We swam over to the coral reef, and I saw for myself what Alex had been talking about back at the dive shop. It was bleached pure white. Which meant death. The water was transparent, and there was no mistaking it. The hot water temperature had scalded all the coral, as it would the rest of the coral all over the Maldives as water temperatures continued to rise over the coming decades.

For a moment I forgot my fear and disappeared into the moment. This was a once-in-a-lifetime experience, and I was lucky to see it. Here, right before me, was firsthand evidence of the effects of climate change.[73] And I wanted to see more. We swam all around the reef for a few more minutes, when Josephine's hand went out again. Out of the corner of my eye I spied a flash of movement in the water. It looked like a big black snake. Of course, it was an eel.

After a smile from Josephine, and a few more minutes paddling around, I said, "OK, Alex, it's time to go back to the boat."

"You're sure?"

"Yes, I've had a ball. But I don't want to press my luck. Can you call them back?"

The boat was nowhere in sight, but I fought a panicky feeling by continuing to snorkel for a few more minutes. The eel was gone now, and the rainbow of fish had reappeared. How many more years remained for others to see this? If the Maldives disappeared, so would any hope of witnessing this bounty of nature. To tell you the truth, I still couldn't get my head around how it was so shallow out in the middle of the ocean. But then again, we'd just learned that this was how the islands had formed, hadn't we? Rising out of the ocean, as a result of sand getting trapped on the coral, forming a natural dredging operation over thousands of years that would form an island. But now all of the coral was dying.[74]

I lifted my head to see that the boat had arrived.

Back onboard, we stopped on the tiny uninhabited island of Odagalla, where we were due for a picnic lunch. Alex and the crew left us to explore the island by ourselves for a while, but this time I didn't mind. It took precisely five minutes to walk around the island, then another tour boat pulled up to find that their uninhabited island was already inhabited by us.

Upon the crew's return after lunch, we headed for the island of Dhaandhoo for more snorkeling. By this point I was all snorkeled out, and there was no more coral. The goal now was sea turtles, in much deeper water near the harbor, which would be cool, but I had a plan of my own. Josephine and Alex slipped over the side of the boat, but I stayed onboard and smiled at the crew.

Suddenly, with their boss gone, one of them could speak English. At that moment a loudspeaker on Dhaandhoo revved up with the midday call to prayer. Because they had to watch the water, my fellow English-speaker explained, they couldn't kneel. But I remained silent while one guy sang his prayers during the music from the loudspeaker. Afterward, we talked.

I asked my new friend if he was originally from the Maldives. He said yes, and the others were too. He said they'd grown up on a neighboring island and had never been outside their atoll. I explained that I was here to study climate change, and he nodded. I then asked, "So, do they teach you everything about climate change in school in the Maldives?"

He shook his head. "Everyone in the Maldives already knows about climate change. The weather has changed so much in my lifetime. There used to be two seasons, now there's only one. We have lots of storms now."

I saw that Josephine and Alex were happily snorkeling and, judging from her voice, she'd just seen a turtle. No need to get very far from the boat this time, as the water was deep, so I stayed seated and continued my conversation with the crew.

"That's terrible," I said. "You've seen climate change firsthand, but people back in the United States don't know all this."

He shook his head curtly, as if wanting to remain polite while still making sure that I understood. "Outside the Maldives, no one cares."

It was heartbreaking.

We spoke for a few more minutes, during which I tried to tell him about the number of people around the world who *did* care about climate change, but they all just looked at me sadly. Throughout our conversation, there was flat affect and not many smiles. Perhaps they understood what no science could teach: that it wasn't just about belief but also giving a damn.

I changed the subject and asked, "So, how many atolls are there in the Maldives?"

This brought a smile and he looked at the other young men, then began to sing a nursery rhyme as he counted on his fingers. Finally he said, "Twenty-six."

At that point I heard Josephine and Alex climbing back into the boat. Apparently she'd seen a number of sea turtles—which were part of an ongoing research project that Alex was doing—and also another shark.

We sat down on the aft to take off our snorkeling gear as the crew ripped through the water as if they were driving the Millennium Falcon, headed straight back to Hadahaa.

We left the island two days later, with another thirty-six-hour journey before us. As we walked with the other tourists down to the dock to catch the boat that would take us back to Kooddoo, the entire staff of the island were lined up to see us off. Everyone was shaking hands and smiling as we walked the gauntlet down the dock. We got into the boat and put on our life preservers, and saw that our fellow tourists seemed as charmed as we were. As the boat pulled away, everyone on the dock started waving. Mile after mile, even as the boat hit cruising speed, the staff on the dock just kept waving. Finally, when I couldn't see them anymore, I imagine they must have stopped, but I couldn't say for sure.

It was both joyful and heartbreaking in precisely equal measure. For I knew that I'd likely never be back. And that, someday, neither would they.

5 Canary in the Coal Mine

The story of climate change is largely one of fossil fuels, such as coal, oil, and natural gas. Everyone understands that cows and other livestock produce methane in their digestive tract, and climate denialists have had a field day with this.[1] But the reality is that the primary sources of greenhouse gas emissions (at least in the United States) are energy production, transportation, and industry, and these are driven primarily by fossil fuels.[2]

In a fascinating story in the *Guardian* newspaper in 2017, I learned that just one hundred companies were responsible for 71 percent of global emissions between 1988 and 2015. Worse, more than half of global industrial emissions are produced by just twenty-five corporate and state-owned entities (see table 5.1).[3]

When I first saw this, a few things jumped out at me. First was that ExxonMobil, Shell, BP, and Chevron were in the top twelve. But wait a minute—weren't those the same companies that had a seminal role in helping the American Petroleum Institute create a campaign to fight the science on climate change, and donated millions of dollars every year to publicity efforts in support of climate denial?[4] Just so. Another thing that jumped out was the enormous influence of coal, which produces about 20 percent of the world's greenhouse gas emissions. Indeed, the number-one producer of greenhouse gases in the world *by far*—accounting for more than the next five entities *combined*—is Chinese coal.

One of the few good environmental stories in the last several years is that the United States has begun to reduce its dependence on coal. Coal use in the US dropped 18 percent in 2019, which means that it ended the decade at less than half the level it started.[5] Still, coal accounts for about 25 percent of all electricity production in the US, even while we transition to

Table 5.1

Top 100 producers and their cumulative greenhouse gas emissions from 1988 to 2015

Count	Company	Percentage of global industrial greenhouse gas emissions
1	China (coal)	14.32%
2	Saudi Arabian Oil Company (Aramco)	4.50%
3	Gazprom OAO	3.91%
4	National Iranaian Oil Co	2.28%
5	ExxonMobil Corp	1.98%
6	Coal India	1.87%
7	Petroleos Mexicanos (Pemex)	1.87%
8	Russia (coal)	1.86%
9	Royal Dutch Shell PLC	1.67%
10	China National Petroleum Corp (CNPC)	1.56%
11	BP PLC	1.53%
12	Chevron Corp	1.31%
13	Petroleos de Venezuela SA (PDVSA)	1.23%
14	Abu Dhabi National Oil Co	1.20%
15	Poland Coal	1.16%
16	Peabody Energy Corp	1.15%
17	Sonatrach SPA	1.00%
18	Kuwait Petroleum Corp	1.00%
19	Total SA	0.95%
20	BHP Billiton Ltd	0.91%
21	ConocoPhillips	0.91%
22	Petroleo Basileiro SA (Petrobras)	0.77%
23	Lukoil OAO	0.75%
24	Rio Tinto	0.75%
25	Niegerian National Petroleum Corp	0.72%

Data from the CDP, *CDP Carbon Majors Report 2017*, cdp.net

other, cleaner fuels.[6] But this must be balanced against the worsening story of coal use in other countries.

> Asia accounts for three-fourths of global coal consumption today. More important, it accounts for more than three-fourths of coal plants that are either under construction or in the planning stages. . . . Indonesia is digging more coal. Vietnam is clearing ground for new coal-fired power plants. Japan, reeling from [the] 2011 nuclear plant disaster, has resurrected coal. The world's juggernaut, though, is China. The country consumes half the world's coal. More than 4.3 million Chinese are employed in the country's coal mines. China has added 40 percent of the world's coal capacity since 2002, a huge increase for just 16 years.[7]

When I read this, I started to wonder about the level of climate denial outside the United States. Maybe the effort of fighting global warming should have more to do with combating climate denial overseas and not just in the US? After a little research, I gave up on this idea because, according to a 2014 survey of climate denial *by country*, China had the fewest deniers.[8] Guess who was number one?[9] Indeed, in his analysis of the survey's results, science writer Chris Mooney noted not only that the US was number one for climate denial but that all three of the worst offenders were English-speaking countries.[10] Clearly, there is no necessary link between climate denial and top polluter status. So it cannot all be blamed on motivated reasoning, ideology, and cognitive dissonance. But what then?[11]

Still, we must look for our keys where the light is. What can we do about the Chinese government's newfound commitment to coal? How about seeking to remove the pretext for international unaccountability that has been provided by America's lack of leadership, which largely *has* been due to climate denial? By doing more to help the US reach its commitment to the goals of the Paris Agreement, might this embarrass China into making more effort toward compliance? Even though China may not be full of deniers, the campaign of denial that has paralyzed the US government has surely had a role in allowing China and other polluting nations to evade scrutiny for their own sins.

And there is the fact that even though China is the number-one producer of greenhouse gas emissions (from all sources), the United States is still number two—and historically we have been the top producer of the type of industrial pollution that *got* the world into this crisis in the first place. Add to this the fact that US emissions (from all sources) still account for 14 percent of carbon dioxide pollution worldwide, and there is plenty to do about climate change even without flying halfway around the world again.[12] The United States is both a major polluter and the excuse for much international delay at a time when we don't have a minute to waste. And one of the main problems is and remains coal.[13]

When I got back from the Maldives, I was eager to use my newly acquired firsthand evidence—and the technique rebuttal strategy from Schmid and Betsch—to try to convince my first science deniers. After doing a bit of research, I learned that Pennsylvania is the third highest coal-producing state in the US, just after Wyoming and West Virginia, so I decided to go there and talk to some coal miners about climate change. This time I didn't anticipate any exotic hotels for my journey, so I arranged to stay with a friend who lives in Pittsburgh and see some folks who ran a nonprofit organization called Hear Yourself Think. Hear Yourself Think aims to encourage more conversation across the political divide, with the goal of helping to break down some of the silos of misinformation and ideology that have exacerbated partisanship and led to gridlock in our current era. David and Erin Ninehouser, the husband-and-wife team that started Hear Yourself Think, offer seminars on how to contend with conservative propaganda that has undermined progressive initiatives such as fair trade and universal health care. But they don't meet partisanship with more partisanship. Instead, they teach us how to have more respectful, engaging, and productive conversations on politically loaded topics.

I couldn't have asked for a better fit with my plan to engage climate deniers. Not only were Dave and Erin former union-affiliated political organizers, but they knew a number of coal miners firsthand, as a result of their canvassing efforts door-to-door for various political causes over the years. As it says on their website, they've "knocked on over 100,000 doors" since 2004 to have difficult conversations with people about progressive causes. In addition, since 2015 they've started going to Trump rallies and filming the results, for use in their training seminars. Before going to Pittsburgh, I watched a few of these encounters, which could be quite shocking. I thought I'd had it tough with the Flat Earthers![14] In any case, Dave and Erin were the perfect folks to help me try to set up a meeting with Pennsylvania coal miners to talk about climate change.

My goal was to choose a setting and format that would encourage respectful and open conversation. I didn't want to be the expert college professor from Boston who flew in to teach everyone a class. Given that, we decided to choose the least adversarial setting we could imagine, which was a diner in either Greene or Washington Counties. This is the heart of Pennsylvania coal country and a place where we could break bread rather than sit in a more formal location like a union hall or library and have a debate. Dave and Erin generously offered to help with publicity and said they'd print up hundreds of flyers to distribute during their canvassing. Erin also called some folks she knew and helped arrange my appearance on an NPR show called *The Allegheny Front* to generate some local interest.[15] My promise was to pay for dinner for anyone who'd come talk to me about their views on climate change. Dave and Erin suggested the Eat'n Park in Washington, Pennsylvania, which had a back room with plenty of tables that we could use for a spirited but private conversation.

As the day drew closer, I was already in Pittsburgh to give some talks at Carnegie Mellon, and crossing my fingers for what lay ahead. I began to get a little nervous. My starting assumption was that most coal miners would probably be climate deniers, based on the famous bit of wisdom from Upton Sinclair that "it's difficult to get a man to understand something when his salary depends on him not understanding it."[16] But I'd also read a piece in the *New York Times* titled "People in Coal Country Worry about the Climate, Too," which gave me a bit of pause. Who was I to prejudge? As the article said, "In this hyperpolarized era [it's] too easy to conflate geography with identity."[17] So I decided to back off my expectations and just hear what people had to say. No more attempts at instant conversion either, like I'd done at Flat Earth. This was about listening and building trust.

I'm glad I approached it this way, because my first two conversations were with coal miners that Dave and Erin introduced me to, who said they wanted to come to the dinner but weren't sure they'd be able to make it. They said it was OK to call, though, so I quickly decided I'd be a fool not to get a jump on things. I put together a list of seven questions but vowed not to do it as a script or interview. Instead, I just wanted to keep the questions in mind and have an easy conversation, to take advantage of this opportunity to talk to folks one-on-one. At the event itself, there might be some polarization, but this was my chance to approach the issue in private, if not in person, and really hear what people thought. Here were my questions:

1. How long have you worked in the mining industry? Have you ever worked in the coal mines?
2. People have a stereotyped view that all coal miners don't believe in climate change. What do you think of that?
3. What's your own opinion on climate change?
4. What other opinions have you heard from people you work with?
5. Do you have family members who have a different view on climate change than you do?
6. What would it take to get you to change your mind on climate change?
7. What do you think of the politics around climate change? Do you think anything will get done?

My first conversation was with a man I'll call Steve (not his real name), who was a coal miner for over thirty years and had been a union rep for United Mine Workers for over four decades. He had retired in 2006. I opened with my question about how people had a stereotyped view that coal miners didn't believe in climate change. Steve was cagey. I'd learned from Dave and Erin that he was a former official in the local Democratic Party, so he knew how to pivot on a question. Steve explained that coal miners were a diverse community. He said that a large number of miners these days were well educated, and that he'd worked with teachers, nurses, etc. He said there was about a 50/50 split between Democrats and Republicans. I asked if this diversity extended to their views on climate change, and he said yes. They had views all over the spectrum. First, there were the Trumpers, who believed that anything on climate change was fake news. Second, there were the folks who said, "I'm a coal miner; I have to be in favor of coal." Third, there were those who were "aware of the environment." Cautiously, I asked where he fell on this spectrum. He said he thought the Earth was in trouble. It was a careful answer, but my sense was that he probably put himself in category three.

I brought up the point of how it was probably difficult for him to hold that view, since it put him at odds with some people in his profession. I said, "How do you square the idea that you were doing something that hurt the Earth, and you knew it, with the idea that you had to do it anyway?" I shared the Upton Sinclair quotation about how it was hard to get someone to believe in something when their salary depended on them not believing it. Wouldn't it be easier if he *didn't* believe in climate change?

Then Steve said something so profound that it changed my perspective on the whole issue. "You've got to understand that coal miners are fatalistic," he said. "The work is a kind of slow death."

Sadly, I'd been aware through Dave and Erin of a twenty-five-year-old mine worker who had died in Washington County on the Friday before Labor Day, when they were planning to meet with some coal miners and their families to canvass for our meetup. My conversation with Steve was less than two weeks after that. I paused to think of the implications of what he'd just said. If a miner was willing to go to work every day and risk their *own* life and health, why would they stop just because of a risk to the Earth or to someone else's health, maybe halfway around the world? That wasn't callous; it was reality. There weren't any other jobs, and they had to feed their family. What did anyone expect them to do?

I asked Steve about the idea of coal miners being aware of the risk to themselves versus risks to the climate, and he said he would have lots of stories for me when we met at the diner. I was overjoyed to have had this chance to talk to him one-on-one, and was already looking forward to more follow-up at the meeting.

My second conversation was with a man I'll call Doug (not his real name), who had worked in the coal industry for forty years and was also active with United Mine Workers. He currently served in government, so he was a public official. Dave and Erin had met him at the Labor Day parade and told him about our meet-up. He was interested in the issue and said he would be happy to speak with me and might even be able to attend the meeting. With Dave and Erin, he had expressed frustration with Trump's rhetoric about protecting coal miners, then doing nothing to back it up. But he was equally frustrated with the "hard-liner" environmentalists, who he felt didn't understand how much of his county depended on the coal industry for jobs.

When I got Doug on the phone, he was just getting ready to go to a local meeting, so we didn't have long to talk. He told me that in his forty years in the coal industry, he was "fortunate" to have been on the surface, working in the control room, and doing welding and processing. But he still had black lung. In the coal industry, he said, dust is everywhere, from extraction to processing. I asked whether everybody who was on the surface felt fortunate, due to the danger in the mines. He said no, that there was more money to be made down in the mines . . . and that all of mining was dangerous.

I turned to my question about how lots of people had a stereotyped view of coal miners. He adamantly agreed, "You got *that* right!" He emphasized the role of technology and remote control in modern coal mining. It wasn't a bunch of guys with pickaxes anymore. I asked his views on climate change. Doug explained that his county was supported by coal and gas. If the coal jobs left, the school system tax base would be decimated. "It scares the hell out of me," he said. There were only thirty kids graduating from the high school already. It was small, and continuing to shrink.

But then he said, "I have grandkids. I've lived my whole life here . . ."

I took this to mean that he fell into Steve's third category of "awareness" about the effects of coal on climate change. But he hadn't said that yet.

Doug continued by telling me that there were lots of things that could make coal emissions better, but they cost money. And who wanted to pay for that? The politicians weren't doing much to help. Finally, he said, "Yes, I do believe in climate change."

But then he immediately said, "But look at China." Their coal industry was much dirtier than the American one. And it accounted for much more of the world's pollution, due to their heavy dependence on coal. (He was absolutely right about that.) But you couldn't just eliminate coal in the US. We'd have brownouts. It would be a danger to national defense. We would have to have a plan, and who's got one?

I mentioned that, in 2016, Trump had said he was in favor of coal and would save the industry. From Doug's point of view, I wanted to know, had he done it?

The answer was a resolute no.

"Trump has closed more coal-fueled power plants than any previous administration."[18]

Doug saw this as a betrayal. He said that his county was Democratic, but that 70 percent of them had voted for Trump.

It was nearing time for Doug to leave for his meeting, so I didn't ask any more questions and just let him talk.

He said that coal miners were intelligent and wanted to make things better, but the solution had to be more than just deciding to eliminate coal. He warned that my event might have some representation from the Center for Coalfield Justice, whom he characterized as radical environmentalists who might make it difficult to have a fair-minded conversation. He reiterated that most people were ignorant of what was going on behind the scenes and that coal miners were not stupid. We had to find a solution that worked for everyone. Then he had to leave for his meeting.

I was ecstatic to have had a chance to talk with two people who had such extensive experience in the industry, and had thought so much about the issues. I was also surprised to find that both of them had said they believed in climate change. But why should I have been? The simple idea that people would deny something just because their livelihood depended on it was a naïve assumption on my part. The issue was much more complex than that. Even if miners realized the impact they were having on the Earth, the more immediate issues for them were jobs and money. It was about family. And they had to balance the risks and reality against the idea that the communities they loved would be gutted if the coal industry disappeared overnight. To their credit, these miners were thinking about this in an infinitely more honest way than the fossil fuel titans that Jane Mayer described in her book, who had been engaging in a campaign of denial all these years.[19] The coal miners I had just spoken with seemed to have much more in common with the kids I met on that boat in the Maldives than they did with the denialists in corporate politics. All were just trying to defend their homes.

I checked in with Erin and Dave and learned that, yes, we could anticipate some representation at the meeting from folks from the Center for Coalfield Justice. And a local reporter was planning to come too. Their canvassing had worked! But what would the results be? They explained that the CFJ folks were activists, but not militant. I shouldn't expect any disruptions. It was surprising to them that miners and local coal supporters always seemed so skittish about engaging with them. Dave and Erin had reached out to CFJ and explained the nature of the event—that it was about talking to neighbors, not trying to convert anybody—and they seemed fine with that. But they wanted to be there.

After things had gone so well during my individual interviews, I decided to try to capture as much of that good spirit as I could at the upcoming dinner. In conversation with Dave and Erin, I decided not to try to get anything on audio or video but just have a talk, for fear of influencing the tone of the event. There was already the risk of polarization, and this event was about trying to break through that. So I busied myself with getting ready for my radio show and looked forward to the event.

The Eat'n Park is a well-known local eatery just off the highway in Washington, Pennsylvania, about twenty miles from the West Virginia border.

I got there early with my friend Andy, a philosopher from Pittsburgh who writes about reason and ideology, and we just tried to relax.[20] When Dave and Erin got there, we checked out the back room, sat down for an iced tea, and waited. They told me they'd canvassed up and down Greene and Washington Counties, all the way to West Virginia. Their flyer was great, and we had done everything we could. Now we had to cross our fingers and hope that people would show up.

Half an hour later, when the dinner was due to start, attendance was sparse. In addition to Andy, Dave, Erin, and me, we had Mike (who had worked forty years in the electrical industry, fifteen of them in a coal plant), Nora (from the Center for Coalfield Justice), Trey (also from CFJ), Nancy (a local who had family members who had worked in steel and coal and said she "could see the issue from both sides" even though she was a "tree hugger"), Zef (a local public school teacher), and Steve (the coal miner I had already spoken to on the phone).[21] There was also a local reporter there, who did not participate in the discussion.

I'd asked Dave to open the meeting because he had so much experience in facilitating difficult conversations, and he did a great job. He talked about how the biggest issue was knowing who to trust. We couldn't all be experts on every subject. And we had to figure out how to overcome fragmentation as a result of social media. So here we all were, meeting face-to-face, to share our views. But no matter how diverse these views might be, the only person who could change our mind was ourselves.

We went around the room for introductions, and then we all had a big laugh. Everyone in the room believed in climate change! Forget the 97 percent of scientists, we had 100 percent consensus in that room. I have to admit that my first feeling was disappointment. Was this representative of the industry? Maybe I should have asked to meet at a union hall, which might have felt more like home turf for the miners. Who would want to go to a meeting at a diner—even with the prospect of a free meal—just to fight about their views? Maybe the climate deniers had all stayed home. Erin and Dave looked a little concerned. They had known I was writing a book about climate denial, and there weren't any deniers in the room. But how could I complain about two generous people, who had canvassed hundreds of miles to turn up coal miners for a meeting with a stranger . . . and they'd come through! My job was to meet coal miners, not climate deniers. I was here to learn what people thought, and maybe I'd just heard it. As I'd read in the *New York Times*, "[the] idea that all miners or all communities bound to extractive industries are hostile to change is simply untrue. Appalachia has a rich history of miners—and their wives—organizing on behalf of their health, jobs and the environment; the mine workers union was founded to protect these principles."[22] This was a complex issue, and we still had a lot to learn from one another. I was grateful.

I asked if we could hear from the coal miners first, and Mike opened the conversation. He said there were various environmental techniques to make coal mining cleaner, but there didn't seem to be much push these days for that to happen. We'd have to go after the "deep pockets" to solve that. Yes, the people involved were part of the problem, but perhaps not *all* of it. If you agreed that we were still dependent on coal, how were we going to fix that?

Steve came next and talked again about the "fatalism" of miners. Your job was to get yourself and your body out of the mine at the end of your shift. That was enough.

Nora (from CFJ) spoke a bit about the "hierarchy of needs." Once my basic needs are met, then I *do* care about the needs of other people. But in a crisis situation (where we're facing economic insecurity and a threat to life and health), maybe that comes later. She said she had an uncle who felt that climate change was not real—that it was not human-caused. But when she brought up the potential effect on clean drinking water, he stopped and thought.

Steve said the whole thing was political. We deal with the issue of climate change in fifteen second sound bites on television. What would it take to change people's minds? How about something comparable to what the 1983 film *The Day After* had done to raise awareness about the dangers of nuclear war? Why weren't there more films coming out of Hollywood about climate change? Short of that, it would take something financial, something personal, to get most people to change their views. And not just their views but their behavior. The question on issues like this is always, "Why should I care?" (What was this? I was hearing the same thing from a coal miner in Pennsylvania that I'd heard 13,000 miles away from a Maldivian fisherman!)

Zef (who had arrived a bit late) said that he taught debate in schools, and that we might be able to change more minds that way as well.

Mike pointed out that this issue had been polarized through the political parties. If we could say just one thing that stuck in a person's mind, to cut through the noise of partisanship, that was progress. Social media is a big part of the problem.

Steve agreed that people don't just *believe* in things anymore, they are *indoctrinated*.

Nancy expanded on this, and felt that this was the goal of political parties. To force us to pick sides. To feel that "someone else should be taking care of this," rather than us.

At this point, I was quite pleased with the conversation so far, and we took off in a free-ranging open discussion, where we all tried to think of some solutions. The idea of meeting one another as humans—as we were doing right now—and planting the seed of doubt so that people could change their minds seemed key. People care about their jobs and their own well-being first. And it's harder to care about something if you can't see it, or don't know anyone who has been affected by it. This is why it's good to have people meet one another. I told a few stories about my trip to the Maldives, and people seemed fascinated. Their circle of caring expanded right before my eyes.

This may not have been the discussion I was anticipating when I'd planned the trip, but it was quite uplifting. The food arrived, and we all continued to talk. Because I had to stop taking notes at this point, all I remember is that we had a cordial and open discussion. But after it was over, my views on climate change have never been the same.

I had heard the same message a few times now, from different people, with different stakes in the issue. There was a distinction between *beliefs* and *values*. The question I had started with was "How do I convince someone to believe something they previously didn't?" But the question I faced now was "How do I convince someone to *care about* something—or someone— they previously didn't?" In order to do more about climate change, maybe the issue wasn't just getting deniers to change their irrational views but to delve a little deeper into understanding how those views were a function of their values, so that we could encourage them to care more about something that would affect us all. That kid on the boat in the Maldives had it dead right. Maybe I'd gone all the way there just to hear him say it: "Outside the Maldives, no one cares." But why not? Because they weren't aware that the issue effected them too? Or that they didn't know anyone from the Maldives? How to get over that barrier—which after all seems to be part of the newest denialist strategy, which has evolved from "I don't believe it" to "Why should we wreck our economy to do something about it?"

What do people get out of denial? What is the role of self-opinion and identity in the hierarchy of needs? At some level, do people know that their beliefs are wrong, but they just can't motivate themselves to change, given the dissonance this would create? But what a brave example I found in *three coal miners*, all of whom understood the reality of global warming and were prepared to admit that in public, even though it conflicted with how they'd made their living. If the ultimate solution to the climate crisis isn't just about changing *beliefs* but also about changing *behavior*, how do we do that? The thing that was keeping those coal miners down in the mines wasn't some radical ideological commitment or motivated belief. It was the economic reality that they needed to feed their family. There was no practical alternative. The coal industry, and the politicians in Washington, had failed them just as surely as it had failed those fishermen in the Maldives. Maybe climate inaction was due to something larger than climate denial. Indeed, maybe the fact that there were climate deniers was only a symptom of a larger problem.

When I left Pittsburgh, I wanted to go back and do more follow-up interviews—perhaps next time at a union hall—and I secured promises from a number of people at the diner to bring some of their climate denier friends next time. But then the COVID-19 crisis hit, and it just wasn't possible. And I also started to wonder . . . what was the point? Even if I could convert some climate deniers over to my side, what good would that do? As I'd seen, there were many people who already *believed* in climate change who were still working in the industry. Did I expect them to go on strike or just walk out?

In fact, I wonder if this doesn't explain the results of the climate denier survey in China. Just because someone is enlightened on the issue of global warming, and believes in the truth of the science, this doesn't mean that emissions will stop. We need something more than belief conversion to make a dent in climate change. But what?

Current Situation

In July 2019, the city of Anchorage, Alaska, hit 90 degrees Fahrenheit for the first time in recorded history.[23] On June 20, 2020, the first day of summer, there was a record-setting 100.4-degree temperature in Verkhoyansk, Siberia.[24] Climate change is everywhere, and it is speeding up. The next IPCC report is anticipated in 2022, and it is widely expected to show that the problem of global warming is even worse than the apocalyptic situation the report described in 2018. Even though public awareness of global warming has been growing in recent years, a quorum of American politicians seem firmly entrenched in denial. And in the absence of leadership, the problem marches on.

We face the reality that awareness isn't enough. That perhaps the best way to solve the climate crisis isn't to talk to climate deniers but to seek political change. In some ways, this might sound easier. Given what I faced at the Flat Earth conference, it doesn't give great hope that we can change science deniers' views—at least on the time scale needed to make a difference for global warming. But imagine if you didn't have to convince someone to give up their denialist beliefs . . . all you had to do was vote them out of office!

But this still requires an effective communications strategy, which means that we have to contend with the propaganda put out by fossil fuel interests. Hard-core global warming opponents—some of whom are in the US Senate—are inundated with disinformation. How can we get them past that? And even if we could, what guarantee would we have that changing minds would lead to action? There are values and personal interests at stake here that go beyond facts and truth. Identity plays a role in framing not just beliefs but also values. What we care about. The things that we are willing to take action for. Don't believe me? Just look at the many other countries—especially China—where climate denial is at its lowest, yet pollution rates are sky high. If overcoming denialism were sufficient for motivating action, why hasn't that worked in China? If the key isn't more arguments and graphs—more words and numbers—then what is?

After the COVID-19 pandemic hit, I began to wonder about the obvious analogies with climate change. Here was a worldwide crisis that would require international cooperation to solve, where our very lives were at stake. Indeed, the only real difference seemed to be the expedited time line. Even though global warming is happening right now, all over the world, it's sometimes difficult to get people to realize that. They say, "Oh, that's in the future," or "I haven't seen it where I live," and move on. They don't care, because in their perception it hasn't yet happened to them. But with COVID-19, the projections were so bad that it promised to affect everyone. So we could at least learn from this how to tackle the analogous problem of climate change, right?

But then, to my amazement, COVID-19 started to be politicized and became the newest form of science denial. I will have much more to say about this in chapter 8, but for now let me simply draw a few parallels that seem relevant to the issue of global warming:

1. If we aren't willing to make the economic sacrifices and other changes necessary to save our own lives in the immediate present, why would we be willing to do so for others in the uncertain future (even though this is false reasoning) about climate change?

2. If we can't summon the political will to engage in global cooperation to fight the pandemic now, where people in every country are already suffering badly, what makes us think we will have the political will to do so for climate change?

3. If special interests can so quickly politicize something like coronavirus—by using the most ridiculous conspiracy theories and partisan nonsense—what hope do we have to unpolarize the "debate" about climate change?[25]

This is a gloomy assessment, yet perhaps the disconnect between public opinion and our crisis in leadership on coronavirus may help us to see a way forward. If we don't have to convince the stubborn minority of coronavirus "hoaxers" in order to solve it—we just need better leadership—perhaps that will work for global warming too.[26]

And, in a weird confluence of events, there is some unabashedly good news for the climate that is linked to coronavirus. During the first few weeks of the pandemic, global emissions plunged an unprecedented 17 percent in early April 2020. According to a United Nations report released before the pandemic (in fall 2019), "global greenhouse gas emissions must begin falling by 7.6% each year beginning in 2020 to avoid the worst effects of climate change."[27] And that's just what's happened in 2020. As a result of working at home, less air travel, less driving, and a total lockdown in some countries, one scientific study estimates that "total emissions for 2020 will probably fall between 4 and 7 percent compared with last year." Another report, by the International Energy Agency, projected an 8 percent drop for 2020.[28] That would meet the UN target! Of course, once the pandemic is over, this is not expected to last. And in order to meet the IPCC target of 2 degrees Celsius by 2050, *we would have to sustain this level of decline every single year between now and 2030* (with further cuts until it reached zero by 2050). And who thinks that—given the massive resistance to economic belt tightening in the US and elsewhere during the pandemic—we could do that?

In the US, there was lobbying to "reopen the economy" less than a month into the first wave of the virus. In the immortal words of Donald Trump, which he tweeted less than two weeks after the first stay-at-home orders were announced, "the cure can't be worse than the disease."[29] More darkly, some were suggesting as early as April 2020 that it was the patriotic duty of some Americans to *die* for the sake of saving the economy.[30] That certainly isn't a bellwether for climate change. If there is so little willingness to make economic sacrifices now to save lives during a pandemic, why would there be more commitment (either now or later) for climate change?

As seems obvious, the crisis here isn't just one of disease and denial. Our very humanity is at stake. In a situation like this, perhaps the key to moving forward is *not* to try to convince the garden variety science denier on the street but to take on the whole system. If climate change (and COVID-19) have been so politicized, why not go after the politicians? They seem to be the only ones who can make big systemic change. And with the public opinion polls on climate change finally showing large public support for some action, why not be more optimistic that something can be done?[31]

There is no question that Republican Party politics is a leading cause of resistance to taking action on climate change. We saw in Mayer's work how the machinery of campaign donations and corporate influence leads to gridlock, but how to break out of this? One sure way is to vote. When evolution deniers overtook the Dover, Pennsylvania, school board and voted in a curriculum that sought to put intelligent design side-by-side with Darwin, this led to a million-dollar lawsuit by parents, and all eight board members were swept out of office during the next election.[32] Instead of trying to convince denialist politicians that they are wrong, maybe we can simply vote better ones into office. Once the 130 current members of Congress who have denied or doubted climate change—like James Inhofe, Ted Cruz, and Mitch McConnell—are out of office, maybe we can have some movement.[33] Once Trump is gone, maybe we can rejoin the Paris Agreement. Yet in the meantime, we shouldn't give up on trying to convince the ones who remain. Whether that means changing their beliefs or their "circle of concern," it all helps with the same effort. Minds need to change for action to take place.

We saw some inspiring examples of Republican belief conversion on climate change in chapter 3, where we heard the stories of Jim Bridenstine, James Cason, and Tomas Regalado, all of whom began to change their minds on climate change once the issue hit home for them, either due to personal experience or listening to people they trusted.[34] Isn't it possible to make use of some of the techniques we examined earlier in this book on the remaining holdouts? Instead of condemning Democratic politicians who "compromise" with the enemy, shouldn't we be happy to see more friendships across the aisle? If we armed more Democratic representatives and senators with the techniques they need for persuasion—and perhaps a few more graphs—might good things happen?

According to one recent study in the *Journal of Experimental Social Psychology*, entitled "Red, White, and Blue Enough to be Green," the persuasive strategy of "moral framing" can make a big difference in making the issue of climate change more palatable to conservatives.[35] By emphasizing the idea that protecting the natural environment was a matter of (1) obeying authority, (2) defending the purity of nature, and (3) demonstrating one's patriotism, there was a statistically significant shift in conservatives' willingness to accept a pro-environmental message. In another study, cited by Dana Nuccitelli in the *Guardian*, it was shown that emphasizing the 97 percent consensus among climate scientists made a bigger difference in convincing conservatives than others. Here is an example where the problem isn't one of simply rejecting facts but of effective science communication that can have a salient influence on conservative attitudes.[36]

Persuading *and* voting. Changing beliefs *and* values. Sharing facts but also trying to enlarge the circle of concern. This is the challenge of our time. The problem of climate change is so large and urgent that it requires all hands on deck. Remember that IPCC report from 2018, which said that we had only twelve short years to try to get ahead of the worst effects of global warming?[37] A more recent consensus—as reported by the BBC—seems to be forming around the idea that "the next 18 months will be critical in dealing with the global heating crisis."[38] It's not just that the 2018 IPCC report said that "global emissions of carbon dioxide must peak by 2020 to keep the planet below 1.5 C," but that if we don't have a global climate plan in place by the end of 2020—and the political leadership to see it through—we probably aren't going to achieve it. According to Hans Joachim Schellnhuber, founder of the Potsdam Climate Institute, "The climate math is brutally clear: While the world can't be healed within the next few years, it may be fatally wounded by negligence until 2020." If we seriously intend to cut global emissions in half by 2030, we cannot wait any longer to begin.

The bad news is that the article I just cited is from July 2019, so the eighteen months are already up. The good news is that, since that article was published, the aforementioned sharp drop in global emissions due to the coronavirus means that we just might make the target for 2020. But beyond that? One hopes that the pandemic will be over as quickly as possible, so that no more lives will be lost. But this means that we must immediately turn our attention back to the lingering crisis of global warming, and we'd better have a plan.

In order to address climate change—whether at the level of individuals or governments—I'm convinced of one thing: we will need to start talking to one another again.[39] In this book, I have focused on the importance of personal, face-to-face engagement as the preferred means of combating science denial. That is because the best way to build trust and respect—and therefore change minds—is through a personal relationship. This is due in large part to the fact that our identity, values, and affect are all involved in belief formation. But isn't the same thing true for deciding what we *care* about? With climate change, we've seen that the best route for achieving a global solution might not be through belief change through individual conversations with climate deniers. Yet, if instead we are trying to make more people care about the issue, the prescription might be the same. If we are trying to change someone's heart or values, the best way to approach them is through personal engagement. People care about people they know. They care about places they've seen. If we can enlarge their circle of concern to include both the Pennsylvania coal miner and the Maldivian fisherman, wouldn't this stand a better chance? It's not wrong to think that, at some level, the process of trying to change someone's beliefs is the same as trying to change what they care about. Perhaps there aren't any ideal argumentative strategies for that, though it's still a good idea to talk. If we remain in our silos, the problem will only get worse.

6 Genetically Modified Organisms: Is There Such a Thing as Liberal Science Denial?

Some have suggested that science denial is by and large a right-wing phenomenon. Examples are not hard to find. We just saw how climate change has been politicized by the Republican Party to become a virtual litmus test of political identity.[1] Another salient example is belief in Darwin's theory of evolution by natural selection, where we find not only a sharp partisan split in polling data but also a thinly veiled public relations campaign by some conservative-friendly evangelical Christians, who are trying to dress up creationism as the new science of intelligent design with the goal of getting it into the science curriculum in American public schools.[2] These are not even close calls: only 27 percent of Republicans think that climate change is a major threat, compared to 84 percent of Democrats.[3] On evolution, only 43 percent of Republicans (compared to 67 percent of Democrats) believe that humans have evolved over time—and the percentage of Republicans who think that has been *shrinking* since the last time this was measured.[4] But does this mean that there are no examples of science denial where the politics are more equivocal?[5] Or even reversed?[6]

Here we must be careful. We've already seen in Lilliana Mason's work that there is a plausible distinction between political ideology and partisan identity. In some cases, what we believe isn't as important as who believes it with us. To some extent, conservative denial of climate change and evolution may be explained by the fact that *this is just what conservatives are expected to believe*, rather than any deep-seated convictions about carbon taxes or how the eye is too complex to have come about by natural selection. So, to the extent that climate change and evolution have already been politicized, it is not surprising to find a partisan split between liberals and conservatives. Once people get the memo about *what* to believe, they can adopt their team's talking points to back up their views.[7]

But is all science denial like this? And, if so, doesn't this allow for the possibility that there could be examples of science denial not only from the political right but also from the left? The salient candidates here—offered by scholars and pundits alike—are anti-vaxx and anti-GMO.[8] In an oft-cited 2013 essay in *Scientific American*, Michael Shermer defends the thesis that politics has the potential to distort science at both ends of the spectrum.[9] More than this, he suggests the provocative idea that—in addition to the well-known problem of Republican science denial—there's also a "liberal war on science." In a later essay, he explains:

> Those on the left are just as skeptical of well-established science when findings clash with their political ideologies, such as with GMOs, nuclear power, genetic engineering and evolutionary psychology—skepticism of the last I call "cognitive creationism" for its endorsement of a blank-slate model of the mind in which natural selection operated on humans only from the neck down.[10]

This looks testable, but remains controversial. Shermer favorably quotes psychologist Ashley Landrum, who explains that "people with more knowledge only accept science when it doesn't conflict with their preexisting beliefs and values. Otherwise, they use that knowledge to more strongly justify their own positions."[11] We already know from Daniel Kahneman's work on cognitive bias that *all of us*—Democrat and Republican, liberal and conservative—have the same cognitive biases that evolved through the process of natural selection over hundreds of thousands of years. Just because someone is left-wing doesn't mean they are immune to the influence of something like confirmation bias or motivated reasoning. Remember Dan Kahan's experiment cited in chapter 2, where liberals *who were good at math* were unable to come up with the right conclusion from a data set when the topic was gun control?

But now to the essential question: is it true to say that there are some areas of liberal science denial? In several of his many works on the topic of conspiracy theories, cognitive bias, and the reasons that people reject science, Stephan Lewandowsky has made the case that there is "little or no evidence for left-wing science denial"[12] and that distrust in science "appears to be concentrated primarily among the political right."[13]

The stakes of even considering this question are high, so it is important to be clear what it might mean to say that liberal science denial exists. Would it be enough to show that there were *some* liberal deniers on *some* scientific topics? That would be too easy, and it has already been established.

Simply to show that 16 percent of Democrats do not think that climate change is a serious issue—or 33 percent have doubts about evolution—is enough to show that there is (some) liberal science denial.[14] But that does not seem to be what Shermer had in mind.[15] At the opposite extreme, must we find some topic on which the rejection of scientific facts or evidence was *exclusively* liberal? That is too stringent. Just as the fact that there are some liberals who deny the reality of global warming does not undermine our conclusion that climate denial is primarily a right-wing phenomenon, one need not find some area where 100 percent of the deniers are liberals to show that there is such a thing as left-wing science denial.

So what are we looking for, then?

How about a case where a majority of deniers are liberals? Or where the driving force behind the argument made against some scientific consensus depends on a core liberal belief? The perils of searching for precisely the right case are outlined well by Chris Mooney, in his continuing engagement with this issue.[16] Mooney's primary concern is (rightly) one of false equivalence. Even if we found some area of science denial that skewed left, this would *not* mean that we had balanced things out against the onslaught of science denial that has come from the right.[17] So if we are going to claim that anti-vaxx or anti-GMO, for instance, are examples of liberal science denial, we had better be prepared to defend the idea not only that more of their adherents are found on the left but that they are motivated by some liberal orthodoxy, in the same way that climate change denial not only bends right but is motivated by conservative views, such as skepticism about government control or an unshakable commitment to free market solutions.[18] Still, as a starting point, even if the overall score card for science denial remained heavily weighted toward the right—which is precisely what Mooney argues in his books *The Republican War on Science* and *The Republican Brain*—might there exist some area of science denial that leaned liberal?[19]

One popular candidate is anti-vaxx. Mooney took up this question in his 2011 essay "The Science of Why We Don't Believe in Science":

> So is there a case study of science denial that largely occupies the political left? Yes: the claim that childhood vaccines are causing an epidemic of autism. Its most famous proponents are an environmentalist (Robert F. Kennedy Jr.) and numerous Hollywood celebrities (most notably Jenny McCarthy and Jim Carrey). The Huffington Post gives a very large megaphone to denialists. And Seth Mnookin, author of the new book *The Panic Virus*, notes that if you want to find vaccine deniers, all you need to do is go hang out at Whole Foods.[20]

In the next paragraph, though, Mooney salves himself with the observation that science denial is "considerably more prominent on the political right" and that "anti-vaccine positions are virtually nonexistent among Democratic officeholders today." Can a view be political if your side's politicians don't use them? In later work, Mooney backs off even further, having apparently blanched at the idea that he might have given aid and comfort to the "false equivalence" crowd, writing articles with titles such as "There's No Such Thing as the Liberal War on Science" and "Stop Pretending that Liberals Are Just as Anti-Science as Conservatives."[21]

What if Mooney is right? Right both that anti-vaxx is an example of science denial that is prominent on the left *and* that this does not justify the conclusion that liberals are just as anti-science as conservatives? Would this still qualify as an example of liberal science denial?[22] That would depend. In the latest available polling data, things seem to have shifted somewhat since Mooney's 2011 article. In a 2014 Pew poll, "34% of Republicans, 33% of independents, and 22% of Democrats believe parents should have final say on vaccination."[23] But that is only one way of measuring vaccine resistance. A 2015 Pew poll found that 12 percent of liberals and 10 percent of conservatives said that vaccines were unsafe.[24] What is the correct way to measure anti-vaxx sentiment? And are these partisan differences large enough to make a fuss over, anyway? These days, anti-vaxx seems not only bipartisan but nonpartisan.[25] Maybe the issue hasn't been politicized at all. Indeed, recent scholarship has found that, to the extent that liberals and conservatives make up roughly equal shares of the anti-vaxx universe, those shares are drawn from the extreme wing of each party.[26] As one commentator put it, "It does not matter what your politics are, the more partisan, the more likely you believe vaccines are harmful."[27] Other research suggests that even where both liberals and conservatives are skeptical about vaccines, this might be for different reasons.[28] What does this say about anti-vaxx as an example of liberal science denial?

Anti-vaxx is a fascinating topic, and squarely in the wheelhouse of any book on science denial. But the politics of anti-vaxx are dauntingly equivocal, especially in the age of coronavirus.[29] And, as previously noted, this topic has been covered quite extensively in recent years, and there are several excellent books already available. Seth Mnookin's previously cited *The Panic Virus* is a good place to start, along with Paul Offit's *Deadly Choices: How the Anti-Vaccine Movement Threatens Us All*. An even more recent, completely accessible account is given in Jonathan Berman's *Anti-vaxxers: How to Challenge a Misinformed Movement*,[30] which examines the core arguments and origins of vaccine resistance. So let's look elsewhere.

What we need is a perfect example. Something where the facts are clear, scientists have reached consensus, *and* the general population rejects these findings largely on liberal ideological grounds. Something not only where the majority of deniers are liberals, but the grounds for denial seems rooted in a left-wing worldview. (As a kicker, it would be nice to show that some of the tactical arguments are the same ones made by conservatives about climate change, which liberals would never tolerate, such as "we need more proof" and "the definitive study hasn't been done yet.")

Instead of anti-vaxx, I propose that we take up an alternative, shamefully neglected topic that seems a perfect candidate for this question: GMOs. After we've learned a bit more about the issues—and talked to a few people—we will return to the political question at the end of the next chapter.

Genetically Modified Organisms

Resistance against genetically modified organisms (GMOs) has been underserved in the science denial literature. Just as, until recently, few were talking about Flat Earth, few talk about GMOs. But the reasons are different. With Flat Earth, it is largely because the numbers have been so small and the claims so outrageous that no one took them seriously. With GMOs, it's more the opposite. Intuitively attractive misinformation is so widespread and rarely challenged that many people actually believe it. Though most GMO opponents haven't studied the underlying science, they claim the results are equivocal. That the experts can't be trusted. That more data have to come in. Sound familiar?

There's another reason that GMO denial is a good choice for this chapter, which is that it represents probably one of the easiest test areas for those who want to talk to a science denier, given that almost everyone knows someone who is against GMOs. I myself have friends and relatives who are driven to red-faced anger at the idea that people are "messing with their food" and believe that without public outcry we would never know what food choices were safe and which were not. I understand the point about pesticides, herbicides, artificial food coloring, antibiotics, and even growth hormones, because there is legitimate scientific research that has shown these to be a potential danger.[31] But GMOs? There have been no credible studies that have shown any risk in consuming them.[32]

"But the studies just haven't been done yet," people will complain. "Remember thalidomide? Things slip through scientific vetting, then they find out later they're unsafe." But we've seen this strategy before. My concern here is not that caution or even skepticism is irrational, but that this issue has now gone beyond risk-aversion to full-blown science denial.[33] It is one thing to say, "Why take the chance when I have a choice?" (though note that anti-vaxxers make the same argument), but it is another to say, "All of the work to produce GMOs has been done by evil corporations that are trying to poison us to make a profit."[34] To argue that there is no *advantage* to eating GMOs (in our wealthy, Western democracies) is one thing, but to argue that GMOs were deliberately created for some malevolent purpose is quite another.[35]

What exactly are GMOs? They are crops that have been molecularly altered to improve their nutrition, heartiness, growth, or resistance to a host of threats.[36] It all started in 1994 with the Flavr Savr tomato (though that particular one stopped being bred in 1997), which was modified to prevent spoilage.[37] Many people do not realize that most of the foods they eat today have been genetically modified.[38] The corn people ate in the eighteenth century looked very different from the corn we eat today. Eighty-five percent of today's corn is a result of artificial selection and genetic alteration.[39] Indeed, there is now a strain of corn that is insect-resistant so that farmers can use *less* insecticide. Genetic alteration virtually saved the papaya industry, which would have been wiped out.[40] The most common crops subject to genetic modification are soybeans, corn, and cotton.[41] But the greatest potential GMO success story is rice.

"Golden rice" was developed by academic researchers in the 1990s to help with the problem of vitamin deficiency and food insecurity.[42] Rice is consumed daily by half the world's population. It is estimated that 250 million children throughout the world have vitamin A deficiency, which can lead to blindness or death.[43] By crossing-breeding rice with a particular gene found in the daffodil, researchers found that the resulting rice was very high in beta-carotene, which is an excellent source of vitamin A. (It also changed the color of the rice from white to yellow, which gave it its distinctive appearance and name.) Golden rice is also more drought-resistant than other strains, a giant leap forward in sustainable agriculture, especially in a world threatened by increased heat and drought due to climate change.

Yet there has been virulent and well-organized resistance to *any* genetically modified foods, including golden rice. Greenpeace in particular has come out against golden rice (for fear that its adoption would pave the way for acceptance of other genetically modified foods).[44] Others have been upset at the idea that a good deal of the research on other GMOs has been done by large agricultural corporations, the biggest of which was Monsanto.

You've probably heard of Monsanto. They were gobbled up by Bayer in 2018 (and the original name was retired), but people still know the legacy: Agent Orange, PCBs, and the long-suspected carcinogenic herbicide Roundup.[45] It was particularly upsetting to some people that one of Monsanto's main GMO products was an herbicide-resistant type of seed that would survive the agricultural use of Roundup to kill adjacent weeds. This enabled farmers to plant rows closer together, but of course it did nothing to assuage the worries of people who (1) did not trust anything that Monsanto did and (2) did not want herbicide on their food. This culminated in the March Against Monsanto (and GMOs) in 2013. It did little, however, to draw attention to the important distinction between saying that GMOs are unsafe to eat versus saying that some GMOs permit the use of more pesticides and herbicides and that *they* are unsafe to eat.

Perhaps Monsanto deserves a certain level of distrust given their corporate history.[46] But to say that *all* of the different ways that GMOs can be created are therefore suspect is to take the argument too far. (Remember, after all, that golden rice was discovered and developed by university researchers, not by Monsanto or any other corporation.)[47] As one commentator put it, it would be "like being against all computer software because you object to the dominant position of Microsoft Office."[48] While it is true that many of the companies doing work on GMOs are agricultural corporations, there is *no scientific evidence* that any GMO products—no matter where they are created—are unsafe.[49] As the American Association for the Advancement of Science (AAAS) put it in a recent statement:

> The science is quite clear: crop improvement by the modern molecular techniques of biotechnology is safe . . . the World Health Organization, the American Medical Association, the U.S. National Academy of Sciences, the British Royal Society, and every other respected organization that has examined the evidence has come to

the same conclusion: consuming foods containing ingredients derived from GM crops is no riskier than consuming the same foods containing ingredients from crop plants modified by conventional plant improvement techniques.[50]

Yet the backlash against GMOs has persisted. In Europe, there is mandatory labeling of GMO products.[51] In the United States and Canada, there is a voluntary labeling campaign for products that do *not* contain GMOs (based on consumer preference) that borders on the ridiculous.[52] I once saw a box of salt that was labeled as a "non-GMO" product, despite the fact that salt is a mineral and does not even have DNA.[53] But this sort of fear-based marketing is perhaps to be expected in a population that is still so ignorant about the subject.

In a 2018 Pew poll, researchers learned that half of respondents felt that GMOs were a great threat to human health.[54] Here it is tempting to compare GMOs to other issues like climate change, where there is a similarly wide gap between public perception and scientific consensus. An earlier Pew poll, in 2015, found that 88 percent of all AAAS members felt that GMOs were safe to eat, whereas only 37 percent of the general public felt they were.[55] As reported in one commentary, this "fifty-one point gap makes this the largest opinion difference between scientists and the public."[56] And, yes, that includes climate change.[57]

Why are people so afraid of GMOs? In some cases, they aren't sure. Is it perhaps that anything "unnatural" in our food pulls a hair trigger in our brains for something to be avoided?[58] Researchers have found that lower levels of knowledge about science are correlated with more resistance to GMOs, and that plain ignorance of what genetic modification really is may be what leads to so much resistance in the first place.[59] In one study out of Oklahoma State University, researchers found that 80 percent of Americans supported mandatory labeling of *foods containing DNA*, despite the fact that *all* foods contain DNA![60] Such broad-based "chemophobia"[61] was not helped when a scientist named Gilles-Éric Séralini published a 2012 study in which he found that rats fed with Monsanto's Roundup Ready Corn developed more tumors and died earlier.[62] By the time this study was retracted due to flawed methodology, a conflict of interest, small sample size, and selection of rats that were known for spontaneously developing tumors as they aged, the study was already out there and had done its damage.[63]

Like other forms of science denial, GMO denial is fed by a healthy dose of conspiracy-based thinking that has little to do with scientific evidence.

The most interesting claims here are made by historian Mark Lynas, *who is a self-confessed one-time anti-GMO activist*. Lynas has now changed sides and started giving talks intended to call out anti-GMO resistance. He writes:

> I think the controversy over GMOs represents one of the greatest science communications failures of the past half-century. Millions, possibly billions, of people have come to believe what is essentially a conspiracy theory, generating fear and misunderstanding about a whole class of technologies on an unprecedently global scale.[64]

Indeed, Lynas wrote a book, *Seeds of Science*, in which he offers a full account of his conversion along with the details and evidence to support his case.[65] As Lynas explains, a few years after his anti-GMO activism, he began to study the scientific literature on climate change, which culminated in the publication of two well-regarded books on global warming, one of which won a prestigious award. His respect for science increased, along with his frustration at science deniers. When the opportunity came to revisit some of his earlier work on GMOs, he was mortified to discover just how little evidence there had been to support them. As he searched frantically for something to justify his earlier views—and couldn't find it—the cognitive dissonance became unbearable. As Lynas explains:

> There is . . . zero evidence that any genetically modified foods in existence today pose a health risk to anyone . . . [but] we cannot criticise global warming skeptics for denying the scientific consensus on climate when we ignore the same consensus on both the safety and the beneficial uses of . . . genetic engineering.[66]

In his book and elsewhere, Lynas has expressed deep feelings of guilt and regret over his prior role in belittling and undermining the work of so many scientists who were trying to address the problem of food scarcity, which may have led to thousands of preventable deaths from malnutrition. In a 2013 speech that went viral on the Internet, Lynas did a full mea culpa in front of a group of farmers, some of whose crops he said he had likely destroyed during his past anti-GMO activism.[67]

This marks one of the most amazing denialist conversions on record, but the method by which it came about was equally remarkable. No one from the "other side" befriended Lynas to patiently give him the facts about GMOs. Instead, through his own work on climate change, he became a different person and began to identify more with scientists. As Lynas put it, the main factor that accounts for his conversion was that "[he] discovered science." More reflectively, he admits that:

> I was probably primed to change my mind on GMOs only because I had begun to shift my loyalty from one group, the greens, to another, the scientists. Receiving the Royal Society Science Books Prize in 2008 I took, rightly or wrongly, as a trophy of affirmation from the scientific community. If I'd been a tribal headhunter this would have been the equivalent of bringing back the scalp of an enemy chief. And it was only when my reputation was threatened—because my writings on GMOs was shown to be perilously unscientific by the sorts of people I now felt aligned with—that I had to seriously reconsider my position. In other words, deep down I probably cared less about actual truth than I did about my reputation for truth within my new scientific tribe. . . . It wasn't so much that I changed my mind, in other words. It was that I changed my tribe.[68]

Lynas here favorably cites Jonathan Haidt's work on how most people hold their beliefs on "moral" grounds, though they try to dress it up in rational language. As we saw in chapter 2, perhaps this is why it is so hard to get people to change their minds on the basis of facts: because beliefs aren't about facts in the first place. Haidt writes: "If you ask people to believe something that violates their intuitions, they will devote their efforts to finding an escape hatch—a reason to doubt your argument or conclusion. They will almost always succeed."[69] Lynas admits that, back when he was an anti-GMO activist, this certainly applied to him. He recounts a story that happened after his "conversion," when he was asked by a genetics professor at Oxford "if there was anything he could have said differently at the time to convince me. I told him I didn't think so. It wasn't that their [scientific] arguments lacked force. Their mistake was to think that their arguments mattered much at all."[70]

In *Seeds of Science*, Lynas tells a familiar story about the origins of GMO denial. As one might guess, it began not with science but with ideology. When genetic engineering—on both animals and plants—was just starting back in the 1970s, there *were* some scientists who expressed general misgivings. These were based, though, not on any experimental evidence but on more global ethical worries about eugenics and what it meant for scientists to rush in where angels feared to tread. Over time, as these concerns died down in the face of actual empirical results, opposition to GMOs morphed to the realm of ideology.[71] Lynas quotes an EarthFirst! activist who told the BBC:

> When I first heard that a company in Berkley was planning to release these [GMO products] in my community, I literally felt a knife go into me. . . . Here once again, for a buck, science, technology and corporations were going to invade my

body with new bacteria that hadn't existed on the planet before. It had already been invaded by smog, by radiation, by toxic chemicals in my food, and I just wasn't going to take it anymore.[72]

Fueled by such moral certainty, opponents of GMOs were not asking for further scientific study; what they wanted was a total ban. And the way to get there was through lawsuits, publicity, and direct action. The latter meant destroying GMO crops when they were still in the field, which is what Lynas took part in. Other activists created a full-press publicity campaign, including newspaper ads in major media outlets that warned about "the perils of globalisation, criticising advanced technology and denouncing the 'genetic roulette' of crop biotechnology."[73] This was incredibly effective, especially in Europe in the 1990s, where people at first had been either generically in favor of GMOs or didn't care (or know) much about the issue. But by the time the scare tactics had spread, "the percentage of the population opposed to GM foods rose by 20 points. . . . In total only a fifth of Western Europeans remained supportive of GM foods."[74] *And this was achieved in the absence of any scientific evidence that suggested GMO foods were unsafe to eat.*[75]

But that isn't really what this was about. It wasn't that anti-GMO activists were piggybacking on some suspicious scientific finding. Instead, the original opposition to GMOs occurred *before* any empirical evidence was available, and continues to this day despite the fact that there is still no evidence they are harmful. But, Lynas argues, the food safety issue was always a bit of a Trojan horse, where scientific facts got twisted in service of a larger "moral" objection to genetic engineering.[76] In conversation with one of his previous comrades in anti-GMO activism, Lynas reports that George Monbiot conceded the point that "it is absolutely true that there's a scientific consensus on GMO safety [but] for me it was all about corporate power, patenting, control, scale and dispossession."[77] Yet even though the opposition to GMOs is largely political, ideological, moral, and theoretical—rather than scientific—the real-world result of GMO denial has been devastating. "Nearly 20 years later . . . there has not been a single approval of a genetically modified crop [in Europe] for domestic cultivation."[78]

Meanwhile, in the United States and other parts of the world, the GMOs that have made it to market have had a surprisingly favorable effect on some of the issues that environmentalists care about deeply. One scientific study showed that GMO technology has *reduced* pesticide use by 37 percent.

Another study estimates that the adoption of GMO crops has decreased greenhouse gas emissions by twenty-six million tons.[79] But what if Greenpeace had succeeded in its goal of achieving a total worldwide ban on all GMO products? The consequences could be devastating. For one thing, we would need more farmland, which would require deforestation and produce more carbon.[80] Indeed, as Lynas point out, if we resisted anything but natural farming technology (circa 1960), we would need a plot of land the size of two South Americas to feed the planet.[81] And the human consequences could be terrible as well. Lynas focuses in particular on the harmful effects that Greenpeace's campaign against golden rice might have already had on starving children. In one searing passage he writes:

> The anti-GMO campaign has . . . undoubtedly led to unnecessary deaths. The best example . . . is the refusal of the Zambian government to allow its starving population to eat imported GMO corn during a severe famine in 2002. Thousands died because the President of Zambia believed the lies of western environmental groups that genetically modified corn provided by the World Food Programme was somehow poisonous.[82]

Surely the word of a reformed zealot like Lynas might be taken with a grain of salt, but his point has been confirmed by other researchers. Anti-GMO ideology is deeply rooted in the idea that—science be damned—GMOs are actually a danger to human health.[83] When we hear free-floating claims that GMO researchers are suppressing their data, that the invention of GMOs was deliberately done to cause food shortages, that it is intended to make our food *more* vulnerable to pests (so that Monsanto can sell more Roundup), this begins to sound like the conspiracy theories we heard from Flat Earthers and anti-vaxxers.[84]

As we have seen, science denial thrives under the conditions of (1) low information, (2) propensity toward conspiracy theories, and (3) lack of trust. All of these are met by those who insist on the dangers of GMOs, despite scientific consensus to the contrary. And, perhaps unsurprisingly, some of the most common anti-GMO arguments fit the science denier script.

Cherry-Picking

One of the favorite techniques of GMO deniers is to raise doubts that there really is a scientific consensus. This is achieved by cherry-picking lists of dissenters, who may or may not have any expertise in this area. The Greenpeace report *Twenty Years of Failure* states that it is a "myth" to think that

GMO foods are safe to eat and claims that "there is no scientific consensus on the safety of GM foods." But, as Lynas argues:

[This] requires extreme selection bias. This is the ultimate in cherry picking. Greenpeace highlights a statement by a small group of dissenters, while ignoring the National Academy of Sciences, the American Association for the Advancement of Science, the Royal Society, the African Academy of Sciences, the European Academies of Science Advisory Council, the French Academy of Science, the American Medical Association, the Union of German Academies of Science and Humanities and numerous others.[85]

Belief in Conspiracy Theories

As we see in Stephan Lewandowsky's work, adherence to conspiracy theories is an essential part of science denial. It should come as no surprise, therefore, that this is true of GMO denial. As Lewandowsky puts it, "Conspiracy theories about genetically modified food (GMO) usually claim that a bio-tech corporation called Monsanto is engaged in a plot to overtake the agriculture industry with poisonous food."[86] Indeed, in his own work, Lynas argues that the entire anti-GMO movement is simply "one big conspiracy theory."[87]

Reliance on Fake Experts or Discredited Research

Here we must be careful. We surely do not want to claim that every scientist who disagrees with the consensus on whether GMO foods are safe to eat is "fake" or that their research has been "discredited." Nonetheless, we must account for how it is that a study like Gilles-Éric Séralini's is still touted as good evidence for the toxicity of GMO foods long after it has been retracted. The parallels to Andrew Wakefield's work on vaccines and autism cannot be avoided. Although there was no evidence of fraud in Séralini's work, the mistakes were numerous.[88] Even so, some anti-GMO activists have wondered whether the retraction was perhaps part of a conspiracy to cover up the truth about GMOs.

Illogical Reasoning

There are numerous logical fallacies in many GMO deniers' arguments. Here are two. First, the idea that if Monsanto is corrupt, then all manufacturers of GMOs must similarly be corrupt. This is called the fallacy of composition, and it is widely taught to first-year students in informal logic.

A second informal fallacy is the slippery slope argument, which basically holds that if you give them an inch, they'll take a mile. We see this used by Second Amendment supporters to push back against *any* sort of gun control, such as, "If you let them ban AR-15 assault rifles, they'll take away your shotguns next." With GMOs, the argument is, "If we let them make golden rice, pretty soon they'll want to make other GMO foods . . . it's a trap!"[89]

Insistence That Science Must Be Perfect

Here the problem is obvious. From Flat Earthers to climate change deniers, we are always hearing that "the crucial experiment hasn't been done yet" or "we need more evidence." The insistence on "proof" for something that someone doesn't want to believe is a hallmark of science denial. With GMOs (and with anti-vaxx) we often hear, "I don't care what the studies have shown so far, something could still go wrong in the future. There's no proof that they are safe." But that is a caricature of how science works.

Does all of this mean that anyone who raises questions about the science that lies *behind* GMOs is automatically a denier? No, it does not. For those who wish to learn more about some of the scientific issues in the GMO debate, there is an outstanding (and accessibly written) book by Sheldon Krimsky called *GMOs Decoded*.[90] He eschews the politics, polls, and cultural arguments that have swirled around the issue and focuses only on the peer-reviewed scientific literature. Krimsky admits upfront that on the narrow question of whether there is a scientific consensus on the safety of genetically modified foods, the answer is yes. He writes:

> In this book, I accept as a starting position that in the United States scientists are largely supportive of the GMOs that currently are planted and consumed. Based on published statements from professional societies and the scientific literature, any concerns over the human health or environmental effects of this new generation of agricultural products have not been any greater than those of traditionally bred crops.[91]

But, as Krimsky points out, this is not the only question. Various methodological, normative, and regulative concerns need to be taken into consideration. For instance, technically it is *not true* to say that there are no harms that can be associated with foods that are the result of molecular breeding. But the question—at least in the US—is whether these foods are *just as safe* as their traditionally bred counterparts.[92] On this question, if the answer is yes, the presumption is that this means GMOs are safe. But can we know this with certainty about each and every GMO product? Of course not, because science cannot answer *any* question with absolute certainty. But if GMOs as a class are just as safe as naturally grown foods, why should they undergo special scrutiny?

The question becomes one of how comfortable we are with anticipating unintended consequences and assessing potential risk. In the US, following guidelines set out by the United Nations and the World Health Organization, oversight of GMOs has largely centered around the question of whether foods that come to us through molecular breeding are *at least as safe* as those from traditional breeding. If a GMO food is found to be "as safe as its conventional counterpart," then it can be considered "substantially equivalent" despite any underlying chemical differences.[93] In Europe, however, the standard is stricter and may require further testing. As Krimsky puts it:

> The starting point for risk assessment differs significantly between the United States and the European Union. The FDA assumes that foods developed by the addition of foreign genes are *generally regarded as safe* (GRAS) . . . unless proven otherwise, whereas in Europe, the designation GRAS has to be demonstrated after testing is complete.[94]

The foods may be the same, but the philosophy of how to handle any underlying uncertainty and risk assessment is different. In the US, a GMO food is considered innocent until proven guilty; in Europe, it is assumed guilty until (nearly) proven innocent. In the US, there are no federally required risk studies. These are instead left to food producers. In Europe, animal testing is required if compositional analysis gives any cause for concern. And even after such tests, all GMO products must be labeled in Europe.

One appreciates here that there may be a basis for legitimate skepticism over the way that GMO foods are handled (at least in the US) that has *nothing to do* with the claim that no scientific study has ever shown that GMO foods are less safe to eat than their traditionally bred counterparts. But if so, why not simply say this, rather than resort to denialist claims about food safety? If the issue is one of caution, rather than evidence of danger, there is no need to engage in conspiracy theories or question the motives of scientists. If we desire a higher standard for testing and regulation of GMO foods in the US, this should not amount to ignoring science.

Krimsky raises the question of what it might mean to say that there are "GMO deniers":

It is too easy to say that one group follows the science and the other group follows an ideology. That leads some observers to embrace the idea of "GMO deniers," referring to people who leave the science behind in favor of an irrational (or groundless) opposition to genetically modified food. But there is a scientific record of studies that support honest skepticism. Also, European and American scientists see the issues and the risks differently, which can explain why their respective regulatory systems are distinct.[95]

Fair enough. Krimsky's book is a fair-minded, evenhanded analysis of both sides of the issue. He points out many times that there is no scientific evidence to support the hypothesis that GMO foods are unsafe. Nonetheless, given worries about unintended consequences and differential modes of risk analysis, this in and of itself is accepted as grounds for some researchers to be skeptical.[96]

But skeptical of what? Of the safety of eating GMOs? This is where skepticism can creep over into denialism. It is one thing to claim that worries about long-term uncertainty, too-lax testing, and industry-friendly regulation give grounds for caution. But to claim that such worries justify outright bans on GMOs invites accusations of denialism. Here the comparison with vaccine denial seems apt. We often find anti-vaxxers making the claim that there are theoretical worries about vaccines too. That the science is too uncertain to justify proceeding with mandatory vaccines, because there are just too many unintended consequences and risks that we don't know about yet.

But the problem is that for such concerns to be scientifically valid—and thus count as skepticism rather than denialism—*they must be backed up by evidence.* And, for GMOs, where is it? With vaccines, there is the Vaccine Adverse Events Reporting System (VAERS), which documents and catalogs the vanishingly small number of "adverse" events so that they can be investigated. But as statisticians know, correlation does not necessarily indicate causation. Just because a child had an adverse reaction near the time they had a vaccine, this does not mean the vaccine caused it. This is why scientists who have access to the VAERS system must investigate. Then they must decide what to do in those rare instances when people *do* have an adverse reaction that can be traced to a vaccine. But even then the question becomes whether this is part of a larger pattern or an isolated incident. With such an important goal as the preservation of public health, even if one child *did* have an adverse (or even catastrophic) reaction to a vaccine, does this mean that all vaccines should be halted? If every vaccine in the country were suspended every time there was a report to VAERS, how many children would die of the measles or whooping cough? The risks must be weighed against the benefits.

With GMOs, a similar standard might be followed. We already know that 250 million children are at risk of vitamin A deficiency around the world, which can be deadly. And nine million people die each year from hunger.[97] Meanwhile, golden rice has sat in limbo for the past twenty years due to theoretical worries about GMOs, based on the "victories" of some nonprofit environmental organizations that have pushed for a ban on all genetically engineered products.[98] How does that risk-benefit analysis look now?

What is the evidence of harm from eating GMOs? As Krimsky admits, there is none. While it is true that there is no VAERS system for GMOs, we do have ample population-based evidence that GMOs are *safe*.

> Hundreds of millions of people have consumed GMOs for over twenty years with no evidence of ill effects and no lawsuits against the GMO manufacturers, even in the United States, which is a litigious nation. If GMOs are a health threat, surely we would have heard by now.[99]

Two decades of anti-GMO activism based only on theoretical worries, with no evidence of any harm, is when skepticism may edge over into denial. Even so, we could (and should) keep looking for adverse effects. But the question remains, what to do in the meantime? GMO denialists would have us ban these products until they can be "proven safe." But how is that possible? Scientists cannot "prove" that GMOs are safe any more than they can prove that vaccines are. Or aspirin. And in the meantime, children are starving.

It is impossible to have proof or certainty in science, and this is a ridiculous standard to uphold for rational belief about empirical topics.[100] Scientific consensus is based not on whether a result is proven but whether it is well warranted by the evidence. Could it nonetheless be true that some GMOs are unsafe? Yes. In science that is always possible. It is one of the hallmarks of scientific reasoning that further evidence may always arrive to overthrow even the most well-regarded theory. But that does not mean that every denier is actually a skeptic, or that it is rational to withhold our judgment until "all of the evidence is in." As we see with climate change or anti-vaxx, there comes a point where skepticism devolves into denial.

Scientific consensus is the gold standard for rational belief.[101] And with GMOs, we surely have that. As Krimsky notes, between 1985 and 2016, the

National Academies of Science, Engineering, and Medicine (NASEM) have issued nine reports on biotechnology. All have come to the same conclusion: "There is no evidence that foods derived from genetically engineered crops pose risks that are qualitatively different than foods produced by conventional breeding methods. Nor is there any evidence that transgenic crops and the foods derived from them are unsafe to eat."[102]

Still, a skeptic might ask, shouldn't we remain concerned about the possibility of long-term risks, such as any possible link between GMOs and cancer? Yes, we should. But the NASEM study provided data to assuage this worry. The UK (where GMOs are rare) and the US (where they are not) have similar rates of cancer. NASEM also reported that "there was no unusual rise in cancer incidence for specific types of cancer in the United States after 1996, when GMOs were first introduced."[103] We are right to be concerned about the *possibility* of a link, but there is no *evidence* of a link, even after exhaustive study. As a comparison, ask yourself how many times Andrew Wakefield's bogus hypothesis about a link between the MMR vaccine and autism must be debunked for vaccine "skepticism" to become vaccine "denial"?

Perhaps Krimsky is right that there are some grounds for skepticism (or shall we call it "a preference for being more vigilant"?) about GMO testing and regulation. But does this mean that there are only GMO skeptics and no such thing as GMO deniers? If so, here is my question: how much evidence would be enough? If you are willing to insist on proof and certainty for GMOs, why not also for vaccines? For evolution? For climate change? If the folks who insist that GMOs are dangerous and unsafe *no matter what the scientific evidence says* are not denialists, is there any such thing as denialism at all?

Following the pattern I've established so far, I wanted to report on some of my conversations with people who are skeptical of GMOs. If possible, I'd prefer to talk to some actual GMO deniers, face-to-face.

My original plan had been to go to my local Whole Foods in the Boston suburb where I live. Why there? Because I know a number of people who shop there for whom devotion to natural foods is almost an obsession. A lot of people think that Whole Foods has a total ban on GMO products, but that simply isn't true. As of today, they don't even require mandatory labeling. In 2013, Whole Foods announced that within five years they would require mandatory labeling on all GMO products, but in 2018 they quietly "paused" that requirement, with no new target date announced.[1] Right now, as it says on their website, their policy is that every product in their stores labeled as organic is guaranteed not to contain GMOs, and if a product in their store is labeled non-GMO, then that must be true.[2] But that doesn't mean that if a product isn't labeled, it doesn't contain any GMOs. Nonetheless, Whole Foods seems to be working hard to have as many of its products as possible carry a non-GMO label, but the truth is that it is nonetheless possible, especially for packaged foods, that there may be some GMO elements.[3]

Yet even though Whole Foods may not have a total ban on GMO products, if you're looking for a population that is sensitive to the issue, there's probably no better place to go. Indeed, some have argued that Whole Foods' commitment to the "natural is better" philosophy goes so far as to veer into pseudoscience.[4] And, especially if I wanted to find GMO deniers who were liberals, I'd likely have my best luck there. As Michael Shermer writes, "Try having a conversation with a liberal progressive about GMOs—genetically modified organisms—in which the words 'Monsanto' and 'profit' are not dropped like syllogistic bombs."[5] I couldn't wait. After going halfway around the world to talk to people in the Maldives, I could *walk* to Whole Foods. As long as I didn't get kicked out, this would be the easiest research I'd done for this book.

Then the COVID-19 pandemic happened, and my plan turned to ashes. Can you imagine walking up to someone in a grocery store these days, even wearing a mask, and trying to strike up a conversation about any topic, let alone food safety? But I was still sold on the idea of having one-on-one conversations, preferably with natural-food devotees. For one thing, I wanted to try to change my orientation from the one I'd had at FEIC 2018, where I was dissatisfied with my success, and even my approach. I'd been too adversarial. And my conversations had been too hit and run. By this point, I'd read Boghossian and Lindsay's book *How to Have Impossible Conversations*, and I was anxious to try out some of their techniques. I needed to have more empathy. I needed to really listen! Not just to hear the other person's evidence, but to make sure they knew that they had been heard. That was the best way to build some trust. And, simply by asking questions, along the way perhaps I could plant some seeds of doubt. Maybe I could even change some minds. But with whom?

Then I realized that maybe the lockdown created an opportunity. Why talk to strangers and try to *build* a personal relationship when I could start one step ahead by having a conversation with someone with whom I already *had* a personal relationship? If trust was an essential feature of changing minds, why not start with someone who already trusted me, and in fact might have expressed some interest in my previous work on climate change and other denialist topics?

Friends in the Time of COVID

I'd known Linda Fox through mutual friends for over thirty years, when we would meet every year for Thanksgiving at our friends' house in Connecticut. Although we didn't see one another very often, we surely had a longstanding relationship, rooted in mutual admiration and trust. She had known my kids since they were born. She liked my wife and I liked her husband. Linda and I might seem like opposites in many ways, but the truth is that we enjoy one another's company and have been sharing views on many topics for years now. Once, when I had a headache the day before Thanksgiving, Linda and some other friends who were staying at the house tried to convince me that the problem was an imbalance in my chakras. She said that what I really needed wasn't ibuprofen but dowsing. Linda identifies herself as both a psychic and a dowser and offered to help me. I'd already taken an ibuprofen, but, in the interest of friendship and a new experience, I agreed to try. I don't remember the specifics of what came next, but soon I began to feel better. I remember the big smile on Linda's face when she said, "See, it's working."

"Maybe," I said.

"Are you feeling better?"

"Well . . . yes."

"See?" she replied. "I told you."

"Well, OK. But I also took an ibuprofen half an hour ago."

She drew up close to my ear and said, "The important thing is that you feel better . . . not everything is a scientific test."

Like I said, she knows me pretty well.

I may only have one gear, but Linda has two. Over the years, my impression is that she has lived her life by both reason and intuition, guided by a deeply felt desire to connect with other people and always do the right thing. You couldn't really find anyone who is more attuned to human needs and eager to reach out and help.

When I decided to call Linda, in the summer of 2020, to ask about her views on GMOs, I already knew a lot about her beliefs on other subjects. I knew that (like me) she was a liberal. I knew that she was firmly against any sort of denial about climate change. And I knew that for the past decade she had run her own artisanal food company called Sumptuous Syrups out of her home in rural Vermont. Every year at Thanksgiving, it's a treat to get a new batch of these, which my wife and I take home to enjoy until we get a new batch the following year. Before my call, I went down to the refrigerator to look at the bottle, then checked the website. There it was: "From our farmers to you—Organic and naturally grown—Concentrated—GF—No GMOs—Free shipping." I was all set.

Linda and I had spoken by phone the previous week, not about GMOs but just to catch up. She knew I was writing a new book, so at the end of our conversation I asked, "Would you be willing to help me if I call back next week, with a little bit of research for my new book?" She eagerly agreed.

I told her it would be about GMOs, and she confessed that she wasn't an expert. But she had her personal views, and I reassured her that was all I wanted. I asked her not to feel compelled to do any research. I just wanted to call back and interview her about her own thoughts—both as a regular person and as a businessperson. She said OK.

The week flew by, and I called her back to talk.

As soon as she picked up the phone, Linda was ready to go on GMOs. In fact, that morning she'd sent me an article she found that "summed up her views pretty well," so I had read it in advance. I made sure she knew, though, that what I was really wanted was *her* views. She didn't have to justify anything. So we dove right in.

Linda said that she'd been in the natural food arena for fifty years! She'd been reading a bit about the subject of GMOs with interest. She said that most people's reaction to a topic like this was either intellectual or emotional. She said she was half Gemini and half Cancer, so hers was both. At this point I honestly had no idea what she was going to say next, but I put away my list of questions and decided to just listen. I knew her to be thoughtful, intelligent, and one of the kindest people I'd ever met. She knew what I was up to, and she would tell me what I needed to know.

Linda said that she was more concerned with the "trait" people were modifying for than the idea of genetic modification itself. Why were they doing it? For what reason? And who was making money off of it? Environmental concerns seemed to be at the top of her list.

At this point she told me a story about herself that I'd never heard before, to emphasize that she was someone who not only *had* beliefs but *lived* them. She said she was so concerned about nuclear power that she and her first husband had lived off the grid for eight years, with no electricity or running water, just to see what it was like and if they could do it. This was in the 1970s, during the height of concern about the nuclear industry throughout the United States, and she wanted to see if it was possible to live with no resources from nuclear energy. I was impressed. Did I even hold my own beliefs that strongly? I'm a fiend about climate change, but I still drive a car. I have air-conditioning. She reminded me that it's about not just beliefs but actions.

Linda said that her views on GMOs came down to two perspectives: personal and business. As a person, her primary concern was influence on the environment. If GMOs were being created to make the food more nutritious that was fine with her. But there was a big distinction between human versus corporate benefit. It shouldn't just be about making money. She volunteered that she did *not* like Monsanto and their product Roundup. She was adamantly opposed to the kind of GMOs that Monsanto was creating, just so they could sell more Roundup. But she had no problem whatsoever with golden rice.

I asked whether she thought that GMOs were safe to eat. She said that she tends not to buy them because it is unknown if they are safe. There are too many questions about it, so she approaches it with caution. "You could eat GMOs," she offered, "but why would you?" But when I asked if she'd ever done it, she said, "Yes."

Her second perspective was that of a business owner. As the owner of a food company based in Vermont, she wanted to have a non-GMO label. The question here was her market. Who were her customers? What would they want? "Is it my job to educate them?" she asked. Probably not. People have different levels of knowledge and different levels of trust, and she wanted to include as wide an audience as possible for her syrups, which meant that she should have a non-GMO product. That was not only good for business, it was the right thing to do to respect her customers' wishes.

I asked whether part of her decision had anything to do with selling her syrups at Whole Foods. She said no, because she wasn't selling them there anyway, not because of GMOs but because of what you had to do to get your product approved to be stocked in their stores. The barriers were simply too high. She preferred to sell online. I then asked if she herself shopped at Whole Foods, and she said rarely. But that was mostly because the store was an hour and a half away from her home. (Did I mention that she lives in a rural area?) She said she preferred to shop locally. She buys organic foods by choice, but it depends on the item. She avoids soy and corn because they were most likely to have GMOs. She said she is a "conscious" shopper, and one of the things she is conscious about is GMOs.

Then she made me laugh. "But if I ate a GMO product, it's not gonna kill me."

She prefers local foods and told me a story about some dairy farmers in her area who had banded together to forgo growth hormones in their milk, and she wanted to support that. This was the kind of issue she cared about because it was local and personal. She knew those farmers. I asked, "But you would still drink milk from other dairies or eat a GMO food?" She repeated her earlier answer: "Yes, but why would you want to?" It was a damned good question, which goes to the heart of the debate. I jotted down a note to ask her later whether that kind of personal choice, though, might have a ripple effect outside her local community (such as those starving kids who couldn't get any golden rice in other countries).[6]

Then I asked her to say a bit more about Monsanto.

"Do you mean the most evil company in the world?" she said.

She went on to talk about how they wanted to control all of the world's food. They were only interested in profit. She said that pollen blows in from one of Monsanto's fields onto an organic farm, and they sue the organic farmer. She also had major concerns about the use of Roundup. She said her neighbor uses Roundup on his fields, which made her especially sensitive to the issue. Does it contaminate the soil? Does it affect the water that gets to her property? What are the unknown effects?

I asked whether she thought that Monsanto GMOs were unsafe to eat.

She didn't answer that directly, but said, "How much would you want to eat, and why would you want to eat it?"

She explained that she tended to be a "centrist" about these things and took precautions. When she goes to a restaurant, she isn't a stickler. She doesn't ask the server whether there are any GMO foods on their menu and doesn't bring up Monsanto. But she said that she does think about how often she eats food whose origin she doesn't know.

At this point she brought up something I'd never heard before, which is that some farmers put Roundup on their wheat the week before it's harvested. "Why would they do that?" she asked. Have they created a problem with food safety, because now the food has Roundup on it? She said she wondered whether all of the people with gluten allergies were really reacting to something else. "It's not that they can't eat gluten . . . maybe they can't eat Roundup." And, she wondered, does Monsanto profit from marketing nongluten products? She said she had no evidence for all this, but it was just one of her "perverse thoughts" that led her to question how her food was grown and how it was processed.[7]

By this point we were getting near the end of our conversation. So far I was *not* getting a denier vibe from Linda. Her approach was cautious and thoughtful. We hadn't talked yet about the scientific consensus—and I wanted to get her reaction to that—but first I needed to establish a couple of things about her views on other scientific topics.

She said that she was *all for* vaccines. She said she'd even go so far as to say that she was pro-vaxx. I already knew she was against nuclear power, but now I asked for her reasons. She said it was (1) that they had no way to deal with the waste and (2) she didn't like the extraction methods, which were not only bad for the Earth but a harm to Indigenous peoples in the countries where the materials came from. On climate change, she said that in addition to the science, as a psychic who had spent some time on a Native American reservation, she could see that it was bad and that the problem of global warming would only grow worse over time.

We were almost done, so I now felt more comfortable telling Linda a bit more about my book, without worrying that I would influence her views. I confessed that I wanted to use our conversation in a chapter on the question of whether there was such a thing as liberal science denial. I said that so much science denial came from the right (about climate change and evolution) that I wondered whether some could also come from the left. She laughed out loud. "Hell, yes," she said. "The hippies have been doing that since the 1960s." We talked a bit about the problem that so many people these days got their beliefs from disinformation campaigns, which led her to say something so profound that I made her repeat it so I could write it down word for word: "We are susceptible to conspiracy theories when we have no reason to trust."

I paused for a moment before asking the obvious question. "So, Linda, do *you* trust scientists?"

"I trust *some* scientists," she said.

She said she always asked herself, "Who's getting paid to do this research?" It's a fair question. We ended with a promise to talk again next week.[8]

My hopes of finding a science denier to convert were dwindling fast. Linda didn't like GMOs, but was she a denier? Not really, no. I turned for advice to one of my closest friends, an environmental biologist I'd known since we were kids. But within minutes of talking to Ted (not his real name) about some of the facts about GMOs, I wondered if maybe I should interview *him*! If I had built up surplus trust with anyone on the planet, he would be on my short list. I came clean and told him that I wanted to interview him for the book. But part of our trust was such that—as a condition of him telling me what he *really* thought about GMOs and not censoring himself—he wasn't quite sure he *wanted* to be interviewed. I offered not to use his real name. We only had a few minutes to get started—and at this point I didn't even know what his considered views were—so we decided to just jump in and see where it took us. We would decide about the book later. He trusted me that much.

In contrast to Linda, this encounter was heated. Ted said that his perspective was influenced by a book he'd read by Jeremy Rifkin, which he wanted to send me. Had I ever heard of Rifkin? Well, yes, as a matter of fact I had. He was the foil in Mark Lynas's book, who had all but started the anti-GMO movement and recruited Greenpeace to the cause. Lynas was savage in his criticism of Rifkin. I kept that to myself for now and just said, "Yes."

Ted's main worry was one of unintended consequences. "This is the way it works with technology," he said. "Nuclear. Oil. You start off thinking it will be fine, then years down the road you learn about the dangers and sometimes there's no turning back." He said, in fact, that this gave him some slight sympathy for anti-vaxxers! (What was this? I'd known the guy for forty years and never heard him say anything like *this* before.)

Next Ted reminded me that he was a *scientist*! He said he understood evolution and how the genome got to be the way it was. It was due to natural selection over a *long* period of time. Each deliberate step was made for a reason, in response to the environment. And with genetic engineering, we were trying to mess with that. We could end up with bad environmental effects and also unsafe organisms. We were dumping Roundup on the soil, and what would that lead to later? We'd all but ruined milkweed, which was a natural habitat for monarch butterflies. And he had concerns about unintended consequences for GMO products themselves. "What if we change the genes in a bacteria and it ends up causing some horrible disease later on?" And what about regulatory agencies that have let hundreds of dangerous pesticides and chemicals persist on the market for years, without adequate safety studies or even after the science shows they are unsafe? He returned to Monsanto and how their history had been one of pursuing short-term profit, not thinking about long-term consequences. Why take that chance?

By now I was pretty excited and didn't want to interrupt his flow, but I needed to ask a crucial question. It seemed to me that Ted's concerns so far were mostly about (1) unintended effects and (2) the environment. But did he think that GMOs were safe to eat?

Ted said he was sure that science probably had lots of information about this and maybe said it was safe, but he wasn't sure. He didn't know it for a fact. He said he was sure the government had standards for food safety, but he didn't trust the government to tell him whether his food was safe. Down the road, how could you be sure? Ten years from now, they might discover something. And even if the products being made now were safe, future ones might not be. Why open this door? They're introducing a new technology that could be used on other things in the future. It could be dangerous.

I nudged the conversation back to the safety of current GMO foods. Ted said he doesn't eat them for ethical and principled reasons. So it doesn't even matter if science said they were safe, he wouldn't eat them anyway. This was for environmental reasons but also a more basic philosophical objection. "Something doesn't sit right with me about messing with natural processes."

He drew an analogy with giving growth hormone to cows. (I swear, I did not bring this into the conversation.) "That's not genetic engineering," he said, "but it is a kind of interference with natural processes. And when they do that, it can lead to bad effects down the road. Would you want to drink that milk?"

I didn't answer, but did at this point make my first real foray into the conversation. I was hoping like hell he would let me use all this for my book.

"Ted, you're a scientist. But your views here are way out of line with scientific opinion. There's a huge consensus that GMOs are safe to eat. The difference between public and scientific opinion here is even greater than the one on climate change."

Immediately, he backed off on the food safety issue. He said he wasn't saying yes and he wasn't saying no. If I needed someone to say they thought that GMOs were unsafe, he wasn't my guy. What he was saying was that he didn't know. And he wouldn't be eating them anyway, so the point was moot. Ted said that in general, he just wasn't comfortable with the idea of GMO foods. At least not yet. Maybe in ten years. But until then, he wouldn't be eating any.

On a hunch, I asked if he shopped at Whole Foods.

He paused briefly. "Yes?"

I already knew he was a liberal. And a scientist! After college Ted had traveled the world doing firsthand ecological research, then came back and got a graduate degree. Now, with his wife, he ran a conservation research institute devoted to saving native species and their habitats. When I donated money for the carbon offsets after my trip to the Maldives, I sent Ted a check to plant a quarter-acre of trees on one of his projects.[9] On climate issues, his beliefs were solidly in the scientific mainstream. But on GMOs?

We closed with my plea for him to take a day and think it over, then let me know whether I could use him for my project. "But," I warned him, "if we talk again, I'm going to try to convince you."

"Well," he said, "maybe I'll end up convincing *you*."

That night a text pinged in the middle of the night saying it was OK. We were on for tomorrow.

The Conversation of My Life

I've got to admit, I was a little nervous for my follow-up conversation with Ted. I'd known the guy for *forty years*, but somehow this seemed higher stakes than any of the conversations I'd had so far with strangers. After joking around for a couple of minutes, I decided to use a strategy out of Boghossian and Lindsay's book.

"OK, Ted, convince me."

Today was different. Instead of a rushed casual conversation with a friend to ask him how he *felt* about GMOs, today I got a cooler head. The scientist was back.

Ted said, "Well, it all depends on what question you're asking me. Would I eat a GMO? Or do I think they're safe to eat?" He picked up on a theme from yesterday and said he didn't like GMOs because of the overall implications of messing with the genetics of our food. He wasn't sure that was safe in the long term. And he was ready with an analogy.

"It's just like invasive species," he said. "When we bring something in, it always seems like a good idea at the time because we're trying to solve a problem. But then it gets out of control. There are always unintended consequences. Look at the mongoose. They introduced it in Hawaii to take care of the rats, but then it started eating *everything*. It totally messed up the local ecology. Now the mongooses have taken over. It's always something you didn't see coming."

"So," he continued, "what happens when they mess with the genetics of a microbe? And then they screw *that* up? Maybe the research on food safety is OK for now, but that doesn't mean there won't be more problems in the future."

I had to admit it was a good point, but I kept my mouth shut and let him continue. He changed the subject to Roundup, and talked about the side effects it had on other species. And how the benefit of that went to fewer and fewer people, and put wealth in fewer hands. There were lots of bad things that could potentially happen for food safety. And the system that regulated this, he felt, was corrupt. It was way too industry-friendly. If something went wrong, would they catch it in time? He summed up: "There's good reason why people don't want to eat GMOs, even if they're shown to be safe."

Now I asked my first real question. "Have you ever eaten a GMO?"

Ted said that he probably had. When he was at a restaurant, how would he know? But he doesn't support the industry, because they can't show it's safe.

Immediately I had a second question. "But how is the position you've just taken not identical to the one taken by anti-vaxxers? Vaccines are 'unnatural.' They can't be 'proven safe.' Are you against those?" I was hoping to get him to expand on what he'd said yesterday about having "some sympathy" for anti-vaxxers.

He said it was a good question, but you always have to balance the benefits with the risks. With vaccines, there's a private risk. You could get sick without them. But there's also a public risk. You could make other people sick too. If there weren't any benefits to having vaccines, no one would do it. But there is a benefit to vaccines, and it outweighs the risks. "But," he was now ready to drive home the point, "there are *no risks* if I *don't* eat a GMO. I'm wealthy enough that I can afford to buy organic. If I were poor and had to eat GMOs to survive, I probably would. But there's no downside to me not eating GMOs."

"But Ted," I jumped in, "how is that not a position that comes from great privilege? There are kids starving in East Asia, and some go blind from vitamin A deficiency because they can't get golden rice. That isn't made by Monsanto; it's university research. But Greenpeace still opposes it. So let me connect the dots between those kids and you. You don't support GMOs and that's fine for you, but if there were enough people like you who only shopped organic and sent checks to Greenpeace, those kids in Asia would starve and go blind. So there *is* a downside to opposing GMOs, just not for you. It's kind of like the anti-vaxx issue you just brought up. You're causing public harm by refusing to support GMOs."

This is the point in a conversation where you're glad you've got trust. The friendship would survive no matter what. I didn't mean to insult him, but in some form or another, Ted and I had been having this same argument for decades.

"I guess you could say the same thing about any technological innovation," Ted said calmly. He brought up the example of fossil fuels. If he doesn't support fossil fuels, someone will lose their job. But does that mean he should support the fossil fuel industry? There are benefits and drawbacks to any technology. With golden rice, of course, there's a clear benefit . . .

He left the point open here, so I decided to push him.

"So do you support golden rice then?"

He shot back his own question. "Would I support coal if it helped the miners keep their jobs?" The implication was no, and he said it was the same with GMOs. There are always unintended consequences. Increased food production seems like a good thing, but ultimately it leads to overpopulation, which is the source of most of the environmental harm to our planet. And it just leads to more starvation later.

"No one wants to talk about it, but it's true," Ted said. "We're getting close to exceeding the carrying capacity of the Earth. With technology, we could keep pushing it. But do we want to do that? Overpopulation is the real environmental danger. And maybe GMOs are just pushing us further down that road."

"Hold on, Mr. Malthus," I said. "So are those kids who can't get the golden rice just going to die?"

He said that some probably would, but the real question was when. If we destroy the planet and outrun our resources, more will die in the long run. And he felt that GMOs made a contribution to that.

I ended the point by saying that this was easy for him to say, because he had money and wouldn't be one of the people who suffered. I wasn't trying to hurt him, just point out the obvious. This was a guy I'd witnessed turn his wallet inside out to give money to the homeless. Who ran up and intervened once when we saw a shoplifter being arrested because he thought it was an attack. Who had devoted his life to sustainable agriculture because he wanted to help the most people possible. But we were at an impasse.

I decided to reframe the conversation and draw a distinction between his environmental concerns and his worries about food safety. Would he say that his objections to GMOs were primarily about the environment and didn't amount to a claim that they were unsafe to eat?

He dealt with the environmental issue first. Ted said that GMOs were causing long-term harm to the environment. This didn't necessarily mean that they were unsafe to eat in the short run, but by supporting the GMO industry he would be causing other—presumably worse—harms down the road.

I nudged him back to the issue of food safety and brought up the scientific consensus I had mentioned yesterday, which had seemed to catch him off-guard. I explained that 88 percent of AAAS members said that GMOs were safe to eat, while only 37 percent of the public thought they were. That was a larger gap between scientific consensus and public opinion than there was on climate change. There was no credible study that had *ever* shown GMOs were a harm to human health. So why didn't he believe that?

He said it was because it was "understandable that people would be suspicious of GMOs." They were "unnatural." It amounted to humans tampering with the food supply. Next he brought up the "precautionary principle" and closed the loop with what he had been saying yesterday. Imagine someone changing the genes around in a bacteria or virus. It could be a nightmare.[10] And now they're doing something unnatural to the way our food has evolved? "They're doing something that's never been done before," he said. "It takes thousands of years for evolution to occur, always in response to an environmental need, but they're swapping genes around in one day. How do they know it's safe?" He explained again that it might be, but he wasn't sure. His trust in government oversight of corporations doing this type of research just wasn't there.

He'd set me up perfectly for my hardest question. "So how is the position you just outlined not the same one that climate deniers use about global warming? They always say, 'We need more studies,' or, 'It hasn't been proven yet.' But science can never *prove* anything. And I know that you know that. All we've got is scientific evidence. But deniers always say, 'It's not enough.' So how much evidence do you *need* about GMOs?"

Ted really liked the question, and rose to the challenge.

He said the whole thing came down to context. With GMOs, you're doing something that has never been done in nature. And you're asking people to be all right with the idea that you're doing something new with their food. But with climate change, you're asking them *not* to do something. You're asking them to *stop*. He said it was the precautionary principle, sort of in reverse. You can't prove that climate change is happening (though the models say it's a million-to-one shot that it isn't), but it's an obvious precaution to stop polluting so much. By contrast, the precautionary principle with GMOs tells you *not* to mess with our food.

I couldn't see him, but I could hear the satisfaction in Ted's voice. It was his best point so far. We were getting toward the end of our conversation, so I asked the question I'd had so much luck with at FEIC 2018 and beyond. "What would it take to change your mind?"

"About what? Whether I'd eat them? Whether I think they're safe?"

"Whatever you like. All of your positions. What evidence would convince you to give up any of the beliefs you've just described to me?"

He said it reminded him a bit of the contemporary debate over nuclear power in the environmental movement. Ted said there was a big controversy in the green community right now over whether they should be supporting nuclear power because it had zero greenhouse gas emissions. So should he support nuclear power? Again, it came down to the issue of risk assessment and short- versus long-term consequences.

"If they proposed to build a nuclear reactor down the street from my house because it would help with climate change, I'd be against it."

I was sure this was true, but I asked him why.

He said the nuclear industry was always trying to show how safe it was, and maybe in large part that was true. But what if something went wrong? Even if there was just a small risk, the result could be so horrible that when you did the analysis, the rational thing was not to support it. And this was his same attitude on GMOs. All of the science could say it was safe, but something bad could happen later. In the end, he said it would be hard to convince him to support GMOs no matter how good the science was, for all of the reasons he'd given.

It felt like we were at a stopping place, so I thanked him and we promised to share some books. He said he'd send me a copy of Rifkin, and I wanted to send him a copy of Lynas. After a few minutes on other matters, Ted came back to my point about the similarity between his position and that of climate change deniers. "That was a good one," he said. "I'm going to have to think about that one a little bit more."

That sounded good to me—and it sounded like we were done.

But, really, we weren't, and I knew we'd revisit this topic for years, long after my book was published.

Ted wants to make the world a better place, and so do I. We both believe in science. But we've been having a fundamental disagreement about "trust in reason" versus "trust in nature" for almost as long as we've been friends.

In our conversation about GMOs, I didn't convince him, and he didn't convince me.

But I think I did prove something: empathy, respect, and listening are the only ways we're ever going to have a chance to change one another's beliefs. The context of trust and mutual respect is the only thing that made this conversation work. Before we hung up, I promised to think more about his points too.

The more I mulled it over, though, the more I doubted that Ted was a denier after all. Was our disagreement over beliefs or something deeper? Values? I didn't want to change his identity, but I did want to change his circle of concern. I wanted him to care more about the suffering of children now than theoretical worries about wider potential harm in the future. And I'm sure he would like to convince me to be more skeptical than I already am, and to have more respect for the hubris of human ingenuity.

So this conversation isn't over.

And I think that is all to the good.

So what is a GMO denier? If you reject the consensus of scientific opinion that says all current GMO foods are safe to eat, is that enough to be considered a denier? I would think so. But what about the important concern over whether future foods will also be safe? There has to come a point where the evidence is compelling, and such theoretical concerns—even while they cannot be disproven—become unreasonable. As a more general point, perhaps the takeaway is this: questioning the consensus (on any scientific topic) does not in and of itself make you a denier. But refusing to believe the scientific consensus *and* being unwilling to say what evidence—short of proof—would be sufficient to get you to change your mind *is* to be a denier.[11] When anti-vaxxers or climate deniers or Flat Earthers insist on *proof*, that is surely unreasonable. Empirical inquiry just doesn't work like that.

Where does that leave us with GMOs? The position that all current GMO foods are safe to eat is supported by overwhelming scientific evidence, and there is really no credible study that suggests otherwise. Is it possible that someone could make an unsafe GMO food at some point in the future? Yes . . . but it is also possible that they might make a killer vaccine. Or a self-crashing plane. Unless someone is uncomfortable with *all* scientific and technical innovation, it seems unreasonable to pick and choose based on suspicion rather than evidence. We need vaccines and air travel, but don't we also need to feed starving children? And, as with the "debate" over climate change, there comes a point where trust has been earned. Where it's unreasonable to doubt anymore. But skepticism must be earned too. To be a skeptic is not merely to doubt everything simply because one can, nor to be rendered catatonic by fear of the unknown; skepticism requires giving trust when the evidence is unexceptionable, even if (as fallibilism requires) we may end up being wrong. Warrant isn't proof, but it's the best that science has to offer. And if you disagree, please consider the question again: what would it take for you to give up your belief that GMOs are dangerous—and how is your position any different than that held by those who are climate deniers or anti-vaxx?

So Is GMO Resistance an Example of Liberal Science Denial?

Even if we are comfortable with the idea that the rejection of current GMO foods as unsafe to eat constitutes science denial, the question remains whether it is an instance of *liberal* science denial. I had two conversations with liberals who had strong feelings on this issue, and neither one turned out to be a straight-up GMO denier. But even if they had, that wouldn't have settled the issue. So let's turn to the empirical literature on this question, which deserves a closer look.

Recall Stephan Lewandowsky's claim that there is "little or no evidence for left-wing science denial"[12] and that distrust in science "appears to be concentrated primarily among the political right."[13] If this is correct, it means that GMOs (and anti-vaxx) are off the table as potential examples of liberal science denial. While there may be some (or even quite a few) left-wing deniers on various topics, this does not quite rise to a level that constitutes liberal science denial, because there are not enough of them to constitute a plurality and/or the ideology behind their views is not drawn primarily from the left. So how can this be measured?

One way is proposed by Lawrence Hamilton, in a study called "Conservative and Liberal Views of Science: Does Trust Depend on Topic?," which Lewandowsky cites favorably throughout his work.[14] Hamilton selects three purported areas of liberal science denial—the safety of nuclear power, vaccinations, and GMOs—and two others that are conservative—climate change and evolution. He then asked one thousand subjects how likely they were to trust scientists for information on these topics. As expected, liberals were more likely than conservatives to say that they trusted scientists for information on climate change and evolution. But then came the surprise. Hamilton found that liberals were also more likely than conservatives to say that they trusted scientists on nuclear safety, vaccinations, and GMOs. He took this as evidence that there were no areas of liberal science denial.[15]

But the study shows no such thing.

First, to measure whether there is an area of liberal science denial, why not focus on *science deniers*, instead of liberals and conservatives? What Hamilton did was interview partisans, then ask them how much they trusted scientists on various topics in science. Instead of looking at science deniers and measuring the percentage of liberals, Hamilton looked at liberals and measured their level of trust in science *as a proxy for denial*. But even if he is right that trust in science is a good proxy for denial (which it is not, as we will see in a minute), what would this even show? Only that a smaller percentage of liberals are deniers on various scientific topics, as compared to conservatives. It does *not* show that a smaller percentage of science deniers are liberals. In short, even if Hamilton is right, it could still be that a plurality, or even a majority, of deniers on any given topic (such as GMOs) were liberals, in just the same way that most climate change deniers are conservatives.

But there is a worse problem. Hamilton uses the question "Do you trust scientists for information on GMOs?" as a proxy for denialist beliefs about GMOs. But this is flawed. Unless he is telling people in advance *what the actual scientific consensus is on GMOs*, how would they have any idea what they were actually trusting scientists about? Given that the level of public knowledge on GMOs is shockingly low (as we saw in the last chapter), how do we know that people who report "trust" in scientists about GMOs have any clue what scientists would say about them? Indeed, according to a recent Pew poll, only 14 percent of the general population understood that *virtually all scientists agree* that GMOs are safe.[16]

A cleaner way to measure whether the liberals were science deniers would have been to ask, "Do you think GMOs are safe?" or "Do you think that scientists agree that GMOs are safe?" In a nutshell, the problem is this: when you ask a layperson, "Do you trust scientists for information on the topic of GMOs?," they might very well say "Yes" because they are thinking, "Yes, scientists are so smart they must know all the evidence which shows that GMOs are not safe for human consumption." To test this, it would have been interesting if Hamilton had asked a parallel follow-up question. After asking, "Do you trust scientists for information on GMOs?," one might immediately ask, "Do you think GMOs are safe?" I'd wager that in many instances the answers to these two questions would not be the same. Yet Hamilton uses the question of trust as a proxy for denial. But without controlling for subjects' level of knowledge, this is meaningless.

Indeed, the flaw here is so great that it swamps the earlier problem. Which is to say that even if Hamilton *had* proceeded to interview partisans about their scientific views—rather than deniers about their political ones—if he had asked subjects directly about their beliefs in GMO safety, rather than their trust in scientists, he likely would not have concluded that GMOs were a right-wing phenomenon. How do we know this? Because there is independent (and more recent) polling data on precisely this question.

A 2015 Pew poll found that 56 percent of liberals and 57 percent of conservatives said that GMOs were unsafe to eat. That's only a one percentage point difference. Is this enough to support the claim that all science denial swings right? Note that this was the same survey that found that 12 percent of liberals and 10 percent of conservatives believed that vaccines were unsafe. So if your answer is "yes" to the GMO question, it will have to be "no" to anti-vaxx. For the sake of consistency, surely two percentage points are greater than one! But of course this is absurd. These numbers are too close to conclude that there is *any* sort of partisan valence on these issues. It seems to me that the numbers are about equally split on the topics of both GMOs and vaccine denial. Yes, this looks more bipartisan than liberal, but there is no denying the fact that *over half of liberals said that GMOs were unsafe to eat.* Now how does it look to say that there is no such thing as liberal science denial?[17]

Unfortunately, Lewandowsky not only cites Hamilton's work favorably, as support for his own skepticism about the existence of liberal science denial, but sometimes uses the issue of trust in science as a proxy for some of his own conclusions about science denial as well. Aside from this, Lewandowsky's work is a paragon of rigor that has done much to advance the question of what science denial is, where it comes from, and how to fight it. Indeed, for the immediate question at hand, we are fortunate that (along with two coauthors) Lewandowsky has conducted an empirical study on precisely this question of whether denial about climate change, vaccines, or GMOs can be explained by one's politics.

In "The Role of Conspiracist Ideation and Worldviews in Predicting Rejection of Science," Stephan Lewandowsky, Giles Gignac, and Klaus Oberauer take up the fascinating question of whether resistance to the scientific consensus—on all three topics above—can be predicted by one's "worldview" (which could consist either of one's political identification as a liberal or conservative or one's commitment—or lack thereof—to free-market ideology).[18] Lewandowsky et al. also assessed subjects' attraction to conspiracy theories. And what they found was amazing. Specifically, subjects' acceptance or rejection of conservative or free-market worldviews strongly predicted their position on climate denial, weakly predicted anti-vaxx, but *did not predict opposition to GMOs at all.*[19] What does this mean? It means that *even if* a plurality of GMO deniers identified as liberal, it wouldn't really be fair to say that this was an example of left-wing science denial, because it wasn't their liberal ideology that had driven them to this rejection of science.[20]

Lewandowsky's paper is detailed analysis, and I have room here for only a few highlights that are most relevant to the question of left-wing science denial. First, it may come as no surprise that embracing conservativism and a free-market ideology were strongly correlated with climate denial. By and large conservatives deny global warming, liberals do not, and the politics line up just as we would expect. But then why wouldn't these worldviews predict whether someone was a denier about vaccines or GMOs? Well, remember that for anti-vaxx it did, but only weakly. There was a positive association for acceptance or rejection of conservativism, but a negative one for free-market ideology. Why might that be? Presumably, some of the anti-vaxxers were libertarians (who would identify as conservatives) who objected to the idea of government intrusion into their personal lives through mandated vaccines, but some of the other deniers were liberals, who were opposed to vaccinations because of mistrust in pharmaceutical companies. As we've seen, people can believe the same thing for different reasons. Given that, Lewandowsky concludes that anti-vaxx isn't a good candidate for liberal science denial, and rightly so.

But on the topic of GMOs, things aren't quite so clear. The worldview questions had no association one way or the other with someone's position on GMOs. This means that even if someone accepted or rejected the conservative label, there was no difference between them and any others in their anti-GMO views. And it was the same for those who rejected a free-market ideology. So much for distrust in big pharmaceutical companies! This is the basis for Lewandowsky's earlier claim that there is "not much evidence" for the view that GMO denial is correlated with liberalism, in comparison to the way that climate denial is correlated with conservativism. In short, neither one's political identity (commitment to conservativism or not) nor political ideology (belief in free markets or not) predicted the rejection of GMOs.

Case closed, right? Not so fast.

Lewandowsky's findings perhaps vindicate the conclusion that *so far* we do not have much evidence for the hypothesis that there is such a thing as liberal science denial—or that GMO denial is a specific case of it—but it does *not* show that the hypothesis has been disproven. Nor does it show that *all* science denial comes from the right. Remember our earlier concern about what it *means* to say that there is such a thing as liberal science denial? Does it mean that there is at least one example of liberal science denial (so it is not all from the right), or does it mean that there is an area where liberal ideology is predominant? Lewandowsky's conclusions support the idea that it is "case unproven" on the second question, but this does not mean that it might not be true that there *is* a case of liberal science denial out there somewhere, nor even—given some methodological concerns—that this might be true of GMOs.

First, we might worry over whether Lewandowky has chosen the right worldview to correlate with anti-GMO beliefs. On climate change, since this issue has been politicized, a question about conservativism should do the trick. But if it's true that GMOs (or vaccines, for that matter) have *not* yet been politicized by the Democratic or Republican Parties in the US, perhaps we should not expect a question about a person's political identity as a liberal or conservative to ring their bell in an online survey. When subjects are asked to agree or disagree with statements such as "The major national media are too left-wing for my taste" or "Socialism has many advantages over capitalism" would that necessarily trigger the issue of GMOs for identity-protective cognition? With the question of free markets, we may be closer to the question of whether one's *ideological* beliefs might predict one's views about science, but then we had better be sure to pick the right worldview! What in the world does free-market ideology have to do with

GMOs? Yes, if someone were dubious about free markets, they might more likely be skeptical of big capitalist corporations (like Monsanto), so this might be expected to predict anti-GMO sentiment. But that is a long way to travel. What does asking subjects about their agreement with statements like "The free market system is likely to promote unsustainable consumption" have to do with GMOs? Perhaps a different worldview question for GMOs—such as "I think that big corporations cannot be trusted to look out for our health and safety"—might have yielded a different answer.

Second, as Lewandowsky admits:

> Although our sample was representative, it may not have included a sufficiently large number of participants at the extreme end of the ideological spectrum. It is therefore possible that small specific groups on the political left do indeed reject scientific findings—such as GM foods or vaccinations—as is suggested by the public rhetoric of spokespersons that are identified as "left-wing."[21]

Yet, as we have already seen, it is *precisely from these ideological extremes that most anti-vaxxers come from.*[22] Could it be the same with GMOs?[23]

Third, as Lewandowsky has stated elsewhere, perhaps it is the case that this question is vexed by:

> the current historical and political context, in which publicly contested scientific findings primarily happen to challenge the worldviews of conservatives rather than liberals. On this account, the laboratory results would lead us to expect that the reverse pattern could be observed if science were to yield evidence contrary to a liberal worldview.[24]

In other words, even if the two examples of anti-vaxx and anti-GMO did not fit the bill, perhaps that is just an historical accident, and we might expect liberals to be just as motivated as conservatives to deny science, if there were ever an example that crossed their own core worldview.[25]

Finally, no matter the experimental findings, there is no denying that on the subject of GMOs specifically, most of the energy and public advocacy to ban them has come from the left. Lewandowsky (along with Joseph Uscinski and Karen Douglas) writes:

> The scientists and organizations committed to the anti-GMO movement in the United States remain largely associated with the left . . . and the movement has had its biggest victories in U.S. states that lean strongly to the left (i.e., GMO labeling laws have been passed in Vermont).[26]

These worries notwithstanding, there is no denying the empirical result. Lewandowsky is correct that so far there has been little evidential support for the idea that GMO denial comes predominantly from the left. The correlations just aren't there.[27]

So what does correlate? In the same paper, Lewandowsky found that while political worldview could not always predict one's anti-science views, there was something that could: belief in conspiracy theories. As Lewandowsky put it, "The two worldview variables do not predict opposition to GM. Conspiracist ideation, by contrast, predicts rejection of all three scientific propositions, albeit to greatly varying extents. Greater endorsement of a diverse set of conspiracy theories predicts opposition to GM foods, vaccinations, and climate science."[28] No matter whether you are a liberal or a conservative, if you're a conspiracy theorist, you are *much* more likely to be a science denier. The science on that is clear.

In some ways this brings us full circle, back from consideration of whether science denial can be explained by one's political ideology to the strategy of the five tropes. Irrespective of *why* someone might resist the science on a given topic, isn't the more pressing issue *how they try to justify those beliefs*? Yes, to some degree. Remember Schmid and Betsch. But remember also the problem of identity-protective cognition. Once a belief threatens someone's identity, they will reach for whatever they can to oppose it. And the only way to overcome this is to talk to them with as much empathy, warmth, and human understanding as you can muster.

Here it's important for any liberals or progressives who have been reading this chapter with a smile on their face to stop feeling smug. Because remember that—according to the best results from contemporary cognitive science—we are *all* subject to the same cognitive biases that can predispose us to denial about scientific and many other forms of beliefs when we find them threatening.[29] And as for conspiracy theories? There are left-wing ones (about Monsanto), as well as right-wing ones (about government scientists). As we saw earlier, we're all susceptible to psychological forces that can prefigure unreasoned disbelief. Science denial is not somebody else's problem. Think of it this way: if anti-scientific beliefs cannot be fully explained by someone's politics, then we're not immune from it based on our own politics either. Whether liberal or conservative, we're all open to the problem of identity-protective cognition, whether the identity we're trying to protect is a political one or not. And it is important to remember that the first interpretation of the "Is there such a thing as liberal science denial?" question still lingers. As Tara Haelle put it in her *Politico* essay "Democrats Have a Problem with Science, Too":

> Cries of false equivalence miss the point. The issue isn't whether the Democrats are anti-science enough to match the anti-science lunacy of Republicans. The point is that any science denialism exists on the left at all.[30]

I didn't propose to discuss the question of whether there was such a thing as liberal science denial to stir the pot and try to politicize things, nor to suggest a type of false equivalence. I did it because I wanted to take seriously (and try to dispel) the common misconception these days that *every* question about facts and truth is political. It's not. Although it's tempting in these post-truth times—when we see serious discussions about facts, proof, evidence, and lying about the economy, environment, immigration, crime, coronavirus, and a host of other topics on our TV news every day—to conclude that the only explanation for disbelief and denial is one's political identity, that's simply not true. There are many other kinds of identity than political. Even if it's true that taking account of identity-protective cognition is the key to creating an effective strategy for pushing back against science deniers about Flat Earth, climate change, anti-vaxx, and GMOs, remember that only *one* of these issues has been politicized. When politicization happens, it can be virulent, but it's important to remember that the team one roots for can be determined by more than just politics.

There are, of course, intriguing questions that could be raised here about parallels between the rejection of truth in general (about the number of people at Trump's inauguration, whether Greenland is for sale, or whether the coronavirus will someday "disappear like a miracle") and the specific issue of science denial. In my book *Post-Truth*, I argued that one of the most important roots of post-truth (which I define as the "political subordination of reality") was sixty years of largely unchecked science denial. This may have provided a blueprint for the kind of politically motivated fact denial we've seen during the Trump era, but that does *not* mean that science denial is explained by politics, nor that we can dispose of it as easily as having an election. Unfortunately, science denial will be with us long after the era of "alternative facts" in Washington, DC has ended. One reason for this is that much of the disinformation being generated about scientific "controversies"—about climate change, anti-vaxx, COVID-19 conspiracies, and even (quite prominently) about GMOs—is being generated through

Russian propaganda efforts aimed at increasing polarization among the populace and eroding trust in government in the United States and other Western democracies.[31]

Surely one's beliefs about any scientific topic *can* be politicized—as we are seeing right now with the coronavirus. Even if they are not inherently political to begin with—or don't tread on some preexisting ideological belief about free markets or individual freedom—with a bit of partisan stagecraft (or foreign interference), we arrive at a place where wearing a mask during a pandemic can become a political statement. As long as we are choosing teams, even empirical beliefs are fair game for the creation of our identity. And that, unfortunately, is what we have seen in the latest example of science denial: the COVID-19 hoaxers.

8 Coronavirus and the Road Ahead

In early 2000, President Thabo Mbeki of South Africa convened a meeting of experts to discuss HIV/AIDS. This was no small matter, for at the time nearly 20 percent of the adult population of South Africa was already infected—the highest rate in the world. After the conference, Mbeki announced that he felt AIDS was caused not by a virus but rather poor immune health, which could be treated with garlic, beetroot, and lemon juice. Hundreds of scientists, in South Africa and elsewhere, begged him to reconsider, but he ignored them. Why did he do this? Because Mbeki believed a conspiracy theory that antiretroviral drugs like AZT were part of a Western plot to poison African citizens. The outcome was predictable. By 2005, almost 900 people a day were dying of AIDS in South Africa. According to one study from the Harvard School of Public Health, there were 365,000 premature deaths between 2000 and 2005 due to Mbeki's denialist beliefs.[1]

Science denial can kill. Especially when in the hands of a national government, denialist beliefs are unusually deadly. Indeed, this is exactly what we are seeing today with coronavirus in the United States.[2] The COVID-19 pandemic is the latest example of science denial. It grew from nothing at the beginning of 2020 and achieved full-blown denial status within months. As such, it provides a real-time test of the hypothesis that all science denial is basically the same. We see the five tropes of science denial on full display in our newspapers and on our televisions every day. The emphasis and character of denialist claims may change from day to day, but the overall effect is obvious. President Trump is a coronavirus denier, and he has infected millions of his followers with denialist beliefs.

Cherry-Picking
"It's just like the flu." "Most people recover." "This is all just a lot of alarmism." By encouraging selective focus on relatively mild cases, COVID-19

deniers downplay the risk of serious complications and death, which led to a fatal delay in creating an effective plan to fight the virus.

Belief in Conspiracy Theories

There are numerous conspiracy theories about COVID-19 that range from its "real cause" or purpose, to who benefits from it, to the idea that it simply does not exist.[3] Here are a few examples: SARS-CoV-2 (the virus that causes COVID-19) was invented in a government lab as part of a biowarfare campaign against the US; it is part of a "deep state" scheme to kill the economy just in time for Trump's reelection; it is part of Bill Gates's plan to depopulate the Earth and implant us with chips to track our movements; it was created by big pharma so they could profit off a vaccine; doctors and scientists are exaggerating the COVID crisis to get more attention for their work; it is actually caused by 5G cell phone towers (which has led to incidents of vandalism in Britain and elsewhere);[4] the CDC is manipulating the death statistics; government science advisor Dr. Anthony Fauci is somehow profiting off the pandemic; and, finally, the whole thing is made up and there are no patients in the hospitals. The latter led to the creation of the hashtag #FilmYourHospital, which resulted in numerous instances of citizen "researchers" storming the waiting rooms of their local hospitals with cell phone cameras, then pronouncing the whole thing a hoax because they didn't see any sick people. (But since when are sick people treated in the waiting area?) There are a plethora of other conspiracy theories about COVID-19—some of which contradict one another—and many people believe more than one.[5] Of course, conspiracy theories tend to pop up during times of crisis—and are not unknown during previous pandemics—which means that although they are not unique to a pandemic, they are characteristic of it.

Reliance on Fake Experts

During an April 23, 2020, news briefing, President Trump relied on the ultimate fake medical expert—himself—to suggest that perhaps COVID could be cured or treated by bringing light or heat "inside the body," or by injection of bleach or other disinfectants.[6] Before this, he had promoted the idea that hydroxychloroquine was a possible cure, despite the fact that the drug has been shown to have no clinical benefit.[7] More recently, Trump praised the work of Dr. Stella Immanuel as "very impressive" and "spectacular" for her claim that hydroxychloroquine is a potential treatment for the coronavirus and her belief that masks are unnecessary.[8] Further investigation revealed that Dr. Immanuel's other medical opinions include the idea that "gynecological problems like cysts and endometriosis are caused by people having sex in their dreams with demons and witches" and that "alien DNA is currently used in medical treatments, and that scientists are cooking up a vaccine to prevent people from being religious."[9]

Illogical Reasoning

On February 27, 2020, President Trump infamously claimed of COVID-19 that "it's going to disappear. One day it's like a miracle, it will disappear."[10] By summer 2020, Trump had grown fond of the claim that the only reason the US was showing more cases was due to more testing. This is provably false, for if it were true, then the percent positivity rate would not have been climbing as well.[11]

Insistence That Science Must Be Perfect

"Why did the scientists change their minds about masks?" "Why don't they have a vaccine yet?" "Why do public health officials keep changing their recommendations about the best public health measures?" The answer, of course, is that scientists learn over time, and change their views accordingly. Science works through a painstaking process of checking one's hypothesis against the evidence in hope of decreasing uncertainty. As they learn more, scientists' beliefs change. But to a denier, any uncertainty by experts is taken as grounds for the credibility of alternative views, whether or not they are warranted by the evidence.

It is of course depressing to see how much misinformation and disinformation is out there, and that it all arose so quickly. But most shocking to see is that so much of it has originated from our political leaders and their allies.[12] The federal response to COVID-19 under the Trump administration consisted of:

- Dumping the Obama-era pandemic playbook and closing the pandemic office[13]
- Downplaying the severity of the virus when it first appeared
- Ignoring scientific advice about the need for a nationwide testing and tracing plan

- Promoting unproven medical advice and cures
- Insisting that the decision to wear a mask should be a personal choice
- Whipping the idea that the states needed to "liberate" from lockdowns
- Pushing states to reopen before they had met CDC guidelines
- Encouraging schools to reopen (and withholding federal funds if they did not)
- Demanding expedited vaccine approval through the FDA before final testing[14]

The bottom line? Like other forms of science denial, COVID-19 denial was manufactured to suit the interests of those who had something to lose if the scientists were right. Trump's virus denial was part of the Republican plan to keep the economy open, despite the threat of more lives lost to the pandemic, to increase their chance of winning the election.

In some ways, none of this is surprising. COVID-19 hits some of the most familiar fault lines in American politics, which were already in place before the pandemic arrived. Anti-pharma, anti-government, and the populist rejection of elites are all at issue. Individual freedom versus governmental control. The supremacy of financial interests over individual caretaking in a capitalist economy. The COVID-19 pandemic was ripe for exploitation by those with political interests. It is nonetheless shocking to see a campaign of science denial more or less directed by the White House. In previous cases, such as climate denial, special interests were at stake that were later politicized. With COVID-19 denial, the special interests have been political from the beginning.

It is important to realize, however, that there are other interested parties who have also been exploiting the pandemic for their own agenda. Some are predictable, like other science deniers. But foreign influence is also afoot.

Outside Interference

From the very beginning of the anti-lockdown rallies, it was clear that anti-vaxxers' interests were at stake. Government mandates over public health measures were bound to be a flashpoint. But there was also the anti-science distrust of expert opinion, belief in conspiracy theories, and insistence on individual choice to boot. This has led to a "cross-pollinization of ideas as these factions get to know each other," which has led some to worry that COVID-19 denial is calling in reinforcements.[15] Another worry is that anti-vaxxers are merely using these protests as an opportunity to put a "fresh coat of paint" on their own form of science denial and remain relevant.[16]

Yet it is not at all clear which way the wind will blow. There has been some speculation that the pandemic might weaken the position of anti-vaxxers, as millions of people around the world clamor for a vaccine. Indeed, there have been some reports of anti-vaxxers changing their minds in the face of this crisis, announcing that if/when a SARS-CoV-2 vaccine becomes available, they will take it.[17] But one wonders whether, in the fullness of time, there might be a case for the opposite result, as the rush to develop a vaccine raises anew the question of whether any vaccine—especially one that has been hurried through testing—would be safe. In August 2020, Russia announced that it would move forward with a vaccine program in October 2020—with virtually no phase III trials.[18] Trump pressed (unsuccessfully) to have a vaccine before the election. This has caused alarm, and may even make some of the previously discredited anti-vaxx hesitancy concerns seem credible. In spring 2020, an AP poll showed that only 50 percent of Americans said they would take the coronavirus vaccine if it were available.[19] The numbers were worse for Republicans, where only 43 percent said they would get the vaccine (whereas 62 percent of Democrats said they would). In an ABC News/ *Washington Post* poll the numbers were similar, with 27 percent of adults saying they *definitely* would not get the vaccine, and half saying they don't trust vaccines in general.[20] If there are enough defectors, COVID-19 may linger indefinitely as we fail to reach the level required for herd immunity.[21]

But COVID-19 denial has been influenced by other ideologies as well. There have been reports that those with far-right views have been turning up at anti-lockdown rallies.[22] When identity is at stake and rebellion is in the air, it can attract strange allies. Indeed, in some ways COVID-19 denial is the ultimate example of what might happen in the future if we can't stop the politicization of science. Everyone with a grievance will hang it on whatever empirical dispute is handy. Distrust is contagious. Will climate change be next?

So far, however, left-right politics has been the driving force behind COVID-19 denial. All across America, division between "maskers" and "anti-maskers" has broken down sharply along partisan lines. According to an NBC News/*Wall Street Journal* poll, the decision to wear a mask in public ended up being a pretty good proxy for presidential preference in the 2020 election. According to the poll, 63 percent of registered voters said they *always* wear a mask in public; among this group Biden led Trump by 40 points. Of the 21 percent who said they *sometimes* wear a mask, Trump led by 32 points. And, perhaps predictably, for the 15 percent of voters who said they *never* wear a mask, Trump led by an incredible 76 percentage points.[23] Is it any wonder that violence has broken out in some places over businesses telling patrons that they must put on a mask?[24] Science denial and political identity have converged in a simple piece of cloth.

Another way that COVID-19 has been politicized is through foreign influence. We already know from previous research that Russia has been responsible for a steady stream of denialist propaganda on climate change,[25] vaccines,[26] and GMOs.[27] Should it come as any surprise that the same is true of COVID-19?[28] According to researchers at Carnegie Mellon University, nearly half of the Twitter accounts spreading misinformation about coronavirus are likely bots. Approximately 82 percent of the fifty most influential retweeters on ending the lockdowns and various COVID-19 conspiracies have been bots.[29] This fits completely with Russian disinformation efforts to exploit existing fault lines in America to sow more discord and division. Some of these efforts can be traced directly to Russian military intelligence, which has used three English-language websites "as part of an ongoing and persistent effort to advance false narratives and cause confusion" during the pandemic.[30] Apparently China has also gotten into the act, and has been trying to induce panic in the US.[31] It is important to remember that the nexus between science denial, disinformation, and politics does not stop at the nation's border.

Social media companies like Facebook, Twitter, and YouTube of course bear some responsibility for the spread of disinformation about COVID-19. Not only are they the preferred platform for foreign propaganda about the coronavirus, they have been responsible for spreading denialist misinformation about other scientific topics for years. According to one February 2020 study out of Brown University, 25 percent of all tweets promoting climate denial come from bots.[32] We've already seen how many Flat Earthers and anti-vaxxers are converted through YouTube videos. But what can be done about the spread of science denial on social media?

During the pandemic, some social media companies stepped up their efforts to combat false information and conspiracy theories about COVID-19.

Prior to the pandemic, CEO Mark Zuckerberg had struggled with the idea of whether it was Facebook's job to police the sort of misinformation that was shared on his site.[33] Famously, in 2019, he said that although he feared the erosion of truth, "I don't think people want to live in a world where you can only say things that tech companies decide are 100 percent true."[34] At that time, the subject of conversation was misleading political ads (which he decided to allow).[35] By the time of the pandemic—at least for misinformation about the coronavirus—Zuckerberg changed his tune. Amid charges that most COVID-19 misinformation had originated on Facebook, the company responded by pointing out that it had removed "hundreds of thousands of pieces of COVID-19-related misinformation," including content that could "lead to imminent harm including posts about false cures, claims that social distancing measures do not work and 5G causes coronavirus."[36] On August 5, 2020, Facebook even removed a post from the Trump campaign that included a clip of Trump falsely claiming that children were "almost immune" to COVID-19.[37] This, of course, raises the question of why Facebook doesn't have a similar policy on climate denial or anti-vaxx misinformation, but at least it's a step in the right direction.[38]

Other companies, like Twitter and YouTube, have also stepped up. By May 2020, Twitter had started to label misleading information about COVID-19.[39] Twitter also sanctioned the Trump campaign's post about children and flagged some of Trump's personal tweets that contained COVID-19 misinformation. YouTube has started to steer people toward credible news sources.[40] Again, for those who care about the role that social media has played in exacerbating science denial in general—not to mention the larger issue of truth itself—it is frustrating that these companies have not been more proactive in trying to combat misinformation and disinformation that will inevitably cause harm. Perhaps the pandemic will open the door to more of these efforts, on other denialist topics, in the future.

Lessons from Coronavirus: Unify and Conquer

One of the most fascinating aspects of the COVID-19 pandemic has been the chance to see what a denialist campaign looks like in real time, and learn what it may teach us about how to fight science denial in general. Many have noted, for instance, the startling parallels between COVID-19 denial and climate denial.[41] In the coronavirus pandemic, we have a microcosm of the threat from global warming: it is an existential threat to the entire planet that portends fairly drastic economic impact and requires worldwide cooperation to address it. If we look at how the world is handling the pandemic, can it give us any insight into how we might handle climate change?

If so, the moral of the story is not uplifting. We have seen foot dragging and resistance not only to the idea that anything is "really" happening but to the idea that it is worth the sacrifice to do anything about it. And this despite the fact that COVID-19 is an immediate threat to our very lives. If we cannot get people to mobilize and do something about a problem that is affecting them personally right now, how are we going to get them to do anything about a threat that they (wrongly) perceive as happening only to other people in far-off places, perhaps a few decades in the future?

The COVID-19 crisis has also revealed quite starkly the enormous importance of money. Economic considerations have had a lamentable influence on public health decisions about what is surely the greatest threat to human health in a hundred years. We hear the slogan "the cure can't be worse than the disease," as if slowing down the economy were worse than hundreds of thousands of preventable deaths. Trump's itch to "reopen America" can be seen as a transparent response to the idea that if people stay at home too long, and slow down the economy, it would be bad for him politically and for the wealthy interests he represents. Trump ally Lieutenant Governor Dan Patrick of Texas actually said that he thought it would be acceptable for older Americans to volunteer to die for the sake of the economy.[42] And if we are willing to do that—to sacrifice hundreds of thousands of lives so that we do not have to endure the economic pain and hardship of job loss and a lower GDP—it does not make me very hopeful that Americans, at least, would be willing to endure the sort of sacrifice to their lifestyle and consumption habits that would be necessary to lower carbon emissions to the point necessary to reach the IPCC's 1.5 degrees Celsius goal.[43]

There is a strong parallel, moreover, between the denialist campaign against COVID-19 and the one we have seen against climate denial as well, albeit on a vastly accelerated time scale.[44] For both the coronavirus and global warming, the denialist position has followed these steps:

- It's not happening.
- It's not our fault.

- It's not as bad as everyone says.
- It would cost too much to fix it.
- We can't do anything about it anyway.[45]

On a brighter note, might the pandemic hold lessons on how to fight climate denial more effectively in the future? According to one article in *Yale Environment 360*, the answer is yes, though the question remains: can we learn it?

> The virus has shown that if you wait until you can see the impact, it is too late to stop it. . . . You have to act in a way that looks disproportionate to what the current reality is, because you have to react to where that exponential growth will take you . . . Covid-19 is climate on warp speed.[46]

But short of an epiphany about the virtues of advance planning or listening to scientists, we seem stuck where we've always been. We need an effective long-term strategy to fight back against science deniers, whatever the issue. So how can we do that?

It is ironic that in the middle of writing a book about the benefits of engaging in face-to-face conversation with science deniers, this newest example of science denial has kept us locked in our houses, consuming information from digital silos that include a heady amount of misinformation and partisan spin. And all of this is happening at great speed. But still—through all of the media, webinars, podcasts, Zoom meetings, socially distanced visits, and family discussions—people have demonstrated a hunger for true and accurate information. Despite the unique challenges of a pandemic, it is instructive to look at some of the things that *have* been working in the fight against COVID-19 denial, with an eye toward how we might more effectively fight science denial on this and other topics in the future.

Here are a few tools that are consonant with our findings so far in this book.

1. Graphs, charts, and tables work   One of the most compelling means to get compliance in mask wearing, social distancing, hand washing, and other public health measures has been the wide availability of statistics from Johns Hopkins University and the Centers for Disease Control (CDC), which are prominently featured in the upper-right corner of virtually every news broadcast in America (even Fox News). This makes clear just how much our actions are costing us. General pleas from doctors and public health officials to follow these measures have been somewhat effective. But what really turns the tide? Looking at a map of the United States and seeing that your own state, county, or city is in a hot zone.

At first, governors in Republican states such as Texas and Florida might have found it easy to dismiss COVID-19 as the "blue flu," since early cases were found primarily in Democratic, or "blue," states like New York and New Jersey. Indeed, there is even evidence that one of the reasons the Trump administration refused to roll out an initial national testing plan is that the early cases were in Democratic states.[47] As southern and midwestern states began to chafe under the initial lockdown restrictions—wanting more "individual freedom" to eschew masks, gather in crowds, and get the economy going again—they embraced Trump's premature "reopening plan," which ignored some of the best public health advice that had allowed New York and other states to "flatten the curve" and slow the spread of the virus.

The results were tragically predictable. A few weeks after reopening, COVID-19 cases shot up in Florida and Texas. Soon these two states became the leading hot spots for the disease. Despite some lingering denial from their governors, the numbers on the graphs just could not be denied anymore. Citizens started to get upset, and trust in governors plummeted. This finally led to better conformity with public health standards (though too late for many). Once it was clear that "red" (Republican) states were subject to the virus too—as clearly demonstrated on every chart and map on every newscast throughout the country—even President Trump and Vice-President Pence decided to put on masks.[48]

2. Emphasizing scientific consensus works Empirical research by Stephan Lewandowsky, John Cook, Sander van der Linden, and others shows that appealing to the fact of a scientific consensus is one of the most compelling ways to get someone to change their mistaken empirical beliefs.[49] Yes, of course there will be those who deny that there is a consensus. But research shows that even deniers—and, notably, especially conservative ones—can be compelled by scientific consensus.[50] The work cited here was done before the coronavirus pandemic, and mostly involved the acceptance of consensus on climate change, but there is no reason to think that this would not also apply to COVID-19 and other forms of science denial.

With COVID-19, we saw this play out in real time through Trump's evolving view on wearing masks, one of the most effective public health measures for fighting the virus. On April 3, 2020, the CDC made its first recommendation to start wearing cloth face masks when out in public. For months, Trump refused to put on a mask. This not only set a bad example but caused confusion among the public over whether public health officials like Dr. Fauci and Dr. Deborah Birx were right that wearing a mask was a necessary tool in taming the virus. On June 20, 2020, Trump insisted on holding a large political rally in Tulsa, Oklahoma, at which masks were optional. A few weeks later, on July 11, the Oklahoma Department of Public Health announced a massive spike in new COVID-19 cases.[51] On July 12, for the first time in public, Trump put on a mask for a visit to Walter Reed Hospital, announcing, "I love masks in the appropriate locations."[52] Later that month, news came that Herman Cain, one of Trump's primary political supporters, who had been at the Oklahoma rally, had died of COVID-19.[53] Pictures of Cain sitting at the rally, in a crowd of people who also were not wearing masks, circulated widely on the internet. Public health officials were unified in their view that, although they could not be sure what might have led to any particular person getting COVID-19, masks were our best possible defense. Trump later said that wearing masks is "patriotic."

3. Personal engagement is powerful   Anecdotal accounts of science deniers who have changed their minds after empathetic conversation with someone they trust are compelling. These days, those conversations might not be face-to-face, but the fact is that personal experience helps to build trust, and the more personal the better. Bluntly put: if someone knows another person who has gotten COVID-19, they are much less likely to be a COVID-19 denier.[54] And this is infinitely more powerful if they are the one who has gotten sick.[55]

In one of the most shocking cases, a thirty-seven-year-old Ohio man named Richard Rose made repeated posts on Facebook claiming that the coronavirus was a hoax. On April 28, 2020, he posted, "Let me make this clear. I'm not buying a fucking mask. I've made it this far by not buying into that damn hype." On July 2, he posted, "This covid shit sucks! I'm so out of breath just sitting here." On July 4, he died, leaving behind a post that said, "When you see me in heaven don't shit yourselves you judgmental pricks." Over the next week, there were a number of posts from Rose's friends expressing shock and grief that he had died. On July 10, a new one came in from a stranger: "Does he still think it's a hoax though?"[56]

In another case, an Arizona woman named Kristin Urquiza wrote an article for the *Washington Post* titled "Governor, My Father's Death Is on Your Hands." In it she told the story of how her father, Mark Anthony Urquiza, an avid Republican and Fox News watcher, believed Governor Ducey and President Trump when they told him he didn't have to live in fear. When the governor lifted restrictions, and Mark Anthony wanted to go out and sing karaoke with friends for the first time in months, his daughter pleaded with him not to go. But he said, "The governor said it was safe. . . . Why would he say that if it wasn't?" His daughter wrote, "A few weeks later, as he was fighting to breathe and terrified that he might die, he told me he felt betrayed."[57]

4. Content rebuttal and technique rebuttal can be effective tools We already know, based on the work of Schmid and Betsch, that it is possible to present science deniers with information that might change their minds. In the case of COVID-19 denial, what information might that be? With content rebuttal, we might share the studies showing that masks are effective. With technique rebuttal, we might point out the problem with conspiracy theory reasoning, either just before or just after they have heard scientific misinformation. Or—if that doesn't work—perhaps we might exploit their predilection for conspiracist thought. Imagine a conversation in which you are talking to a COVID-19 denier (which means that the person is already predisposed to believe in conspiracy theories), and you share the fact that Russia and China are engaged in a massive disinformation campaign on social media in support of the idea that the coronavirus is a "hoax" and we need to "liberate" ourselves from lockdown restrictions. This is not a conspiracy theory, it is a real live conspiracy! Might this not appeal to them? I've cited some sources earlier in this chapter that you can print out and hand to them. Some even include graphs and charts. You might encourage your interlocutor to think about who is benefiting from all of the polarization and division in the US. Doesn't that make them even a little bit suspicious? If all else fails, suggest that they do their own research. It is perverse, but it just might work!

I wish such tactics weren't necessary. If we had better political leadership at both the state and federal level, science denial might not be such a problem. But don't the rest of us bear some responsibility too? How tempting it is to shut out people who disagree with us and dismiss them as stupid—or change the channel because we can't even stand to see where they are getting their information—as we prefer to talk only to people who already agree with us. My message in this book is simple: we need to start talking to one another again, especially to those with whom we disagree. But we have to be smart about how we do it. Simply sharing information does not work. And insulting people or shaming them for their beliefs *definitely* does not work. If our goal is actually to convince someone to give up their denialist beliefs, we have to approach such conversations with as much empathy and respect as we can, toward the goal of building trust and rapport so that our companion might actually listen.

As I sat in my house writing this book (as a result of an underlying medical condition that made it dangerous for me to risk any exposure to SARS-Cov-2), I lamented the fact that I could not get out there and talk to anti-maskers and other COVID-19 deniers to do more firsthand research. Then in July 2020 a story appeared in the *New York Times* with the provocative title "How to Actually Talk to Anti-Maskers."[58] It felt like someone was reading my mind.

In this article, Charlie Warzel makes the provocative claim that shaming and stigma are exactly the wrong way to get anti-maskers or other COVID-19 deniers to change their beliefs or behaviors, especially in an atmosphere of distrust. This is particularly true on an issue like the coronavirus, where knowledge is evolving rapidly. Warzel starts with an analogy about the Ebola virus crisis. He writes:

> As the Ebola epidemic raged in 2014, some West Africans resisted public health guidance. Some hid their symptoms or continued practicing burial rituals—like washing the bodies of their dead loves ones—despite the risk of infection. Others spread conspiracies claiming the virus was sent by Westerners or suggested it was all a hoax. . . . So the World Health Organization sent Cheikh Niang, a Senegalese medical anthropologist, and his team to figure out what was going on. For six hours, Dr. Niang visited people . . . inside their homes. He wasn't there to lecture. Residents asked him to write down their stories. When they finished, Dr. Niang finally spoke. "I said, 'I heard you,' he told me recently over the phone from Senegal. 'I want to help. But we still have an epidemic spreading and we need your help, too. We need to take your temperature and we need to trace this virus.' And they agreed. They trusted us."[59]

As Warzel points out, the people weren't necessarily selfish or anti-science. They just felt scared and stripped of their dignity. They needed someone to respect and listen to them—which in turn built trust—and as a result that trust was returned.

Contrast this with the situation involving anti-maskers and other deniers around the coronavirus pandemic in the US. The political division and partisan context seems intractable, but is it? As Warzel points out, most people—even Republicans—still trust science.[60] So what is the problem? Perhaps it lies not just with COVID-19 deniers but with the way we have been communicating with them. Consider the fact that this is a *novel* coronavirus. We've never seen it before, which means that we don't know everything about it. This surely accounts for the fact that scientists are learning more over time, which means that their advice and recommendations must occasionally change, sometimes radically. Recall that, up until mid-April 2020, the World Health Organization and other experts were saying it was not necessary to wear a mask in public. Then all of a sudden, they changed their recommendation. Does that suddenly make everyone who didn't want to wear a mask irrational?

The problem is that this kind of reversal opens the door for a failure of trust. Unless it is communicated perfectly, people become suspicious of why the message has changed. Those who are cognizant of the way science works understand that there is always some uncertainty behind any scientific pronouncement, and in fact the hallmark of science is that it cares about evidence and learns over time, which can lead to radical overthrow of one theory for another. But does the public understand this? Not necessarily. And, in an atmosphere of distrust, perhaps scientists and public health officials were reluctant to accept this and approach the topic with the humility and transparency it required.

Although it may sound illogical, admitting uncertainty can actually *increase* trust. Saying you don't know something (and why) can allay suspicion and build credibility for later, when you do. And lying to someone—for instance, by saying that masks are 100 percent effective, or that any vaccine is guaranteed to be safe—is exactly the wrong tactic. When scientists do that, any chink in the armor is ripe for later exploitation, and deniers will use it as an excuse not to believe anything further.

Did this happen with public health orders during the earlier parts of the COVID crisis? As Warzel points out, it almost certainly did. Here he quotes Dr. Ranu Dhillon, a Harvard physician who advised the president of Guinea during the Ebola crisis: "All advice ends up binary. . . . It's absolutely one way or absolutely the other way, when it should be shades of gray. It happened with the World Health Organization and denying the asymptomatic transmission early on. It happened with masks. And it happened with states reopening."

Warzel goes on to explain:

> Given that this is a novel coronavirus, we're learning on the fly—things that are true one day might need to be revised the next. Dr. Dhillon suggests that in their desire to be authoritative, public health experts have eroded trust by not accurately communicating uncertainty and by being stubborn about correcting the record when our understanding evolved. . . . "My perception is that public health officials were hesitant to be too vocal about the outdoors being safe," Dr. Dhillon said. This was an example of experts hedging language to make sure the public didn't over- or underreact. But it's a gamble, trying to control how others receive your message. Dr. Dhillon argued that the same thing happened with masks, where officials and institutions were hesitant to suggest them early on because of supply chain issues. It's possible that the reversal opened the door for the culture war we're now experiencing around face coverings.[61]

Once you've made a confident pronouncement, then taken it back, it's too late. Trust is gone. The time to share qualifiers or express uncertainty is from the very beginning, even if your intentions are pure and you just want to keep people as safe as possible.

I have long held that one of the greatest weapons we have to fight back against science denial is to *embrace uncertainty* as a strength rather than a weakness of science.[62] If scientists always pretend to know the answers— even when they don't—is it any wonder that denialists find quarter with suspicion and lack of trust? As Warzel puts it in his article, "You cannot force public trust; you have to earn it by being humble and transparent, and by listening."

And perhaps this is the bottom line for what has gone wrong so far in the battle against COVID-19 denial. Yes, there is plenty of blame to go around for media outlets who either hyped or downplayed scientific information (and did not let scientists speak for themselves), for politicians who fomented division and promoted disinformation, and even for the public, who were too gullible to change the channel or listen to anyone else once they got their marching orders from their favorite media outlet and heard what they were "supposed" to believe. But some of the blame also rests with communication failure by scientists, physicians, and public health officials who dictated what must be done, rather than explain the evidence or process of reasoning behind it, based on an expectation of authority and infallibility from a bygone era. Given this environment, as the public health recommendations changed, sometimes day to day, it is any wonder that some people were suspicions? Again, it is almost impossible to fathom how difficult it must be for scientists and others to share the latest true and accurate information—while qualifying their remarks with unfailing good faith and humility—when it is so much easier simply to tell us what to do so that more lives will be saved. Worse yet, in an environment in which scientists were assaulted with a steady barrage of slander and political spin—inflamed by the media, who were often trying to emphasize partisan controversy—is it any wonder that some public health officials just wanted us to follow their latest recommendations and leave the uncertainty to them? But that is the challenge of effective science communication, not just for scientists but for all of us who care about it. We must fight the battle we are actually in rather than the one we wish we were in. Science is in a legitimacy crisis these days. This is the fault not of scientists but of media, politics, education, and a broader culture of distrust. But if scientists are not prepared to stand up for science—by sharing its results and modeling its core values of openness, transparency, humility, and a reticence to overstate one's conclusions—who will?

The conclusion Warzel reaches about the pandemic is consonant with the one I have been recommending throughout this book for fighting Flat Earth, climate denial, GMO denial, and anti-vaxx. The most effective way to talk to a science denier is to try to build trust through direct personal engagement, showing humility and respect, while demonstrating transparency and openness about how science works. If people can be educated about the latest public health measures, why not also about the processes of science? But if we do not do this, the next crisis may be over trust in the coronavirus vaccine.[63] Warzel quotes Dr. Tom Frieden, an infectious disease expert and former director of the CDC:

> What worries me most right now is that the distrust we're seeing today will happen with vaccines. There's already a huge amount of distrust in vaccines. We've got this scarily named Operation Warp Speed program for them. That kind of naming is just not the way to convince someone to put a needle in their arm. It runs the real risk going forward that whether or not the government cuts any corners on the construction of a vaccine, there will be a perception it has, unless we have very open and transparent communication about it.[64]

I hope the day will come when SARS-CoV-2 has disappeared (though certainly not "like a miracle"), and along with it COVID-19 denial might disappear as well. I wish we could say that for all of the other forms of science denial too—most importantly for climate denial—which will likely be with us for some time to come. Even after the pandemic, we will still have a lot to figure out. But I hope to have offered a bit more insight and experience about the most effective approach and tools to fight back against the remaining science deniers.[65]

The Cure for Post-Truth

In my book *Post-Truth*, I made the case that today's "political subordination of reality" has strong roots in sixty years of largely unchecked science denial, which started with the tobacco companies that manufactured doubt over the link between cigarettes and lung cancer, right up through global warming. But something incredible has happened in the last few years, which is that political polarization has made the problem of science denial even worse. Post-truth and science denial now seem locked in a self-stoking feedback loop. Disagreements about empirical facts and political values have in some cases merged.

Perhaps this means that the solution for science denial and the larger problem of politically motivated reality denial might be the same too. We must begin to talk to one another again. If you were trying to convince someone to change their political or ideological beliefs, how would you do it? Probably not by trying to fill their information deficit. (And certainly not by yelling or insulting them.) You might mention some facts (if you could get agreement on what those were), but for the most part you would probably try to appeal to a common set of values. To a shared sense of identity. And the only way to do *that* is through trying to build a personal relationship that increases trust.

There is no reason we cannot do this with science. We need to teach people not just the facts of science but also its values, and how those values inform the processes by which science makes its discoveries. One of the main problems with science communication begins with science education. When I was in elementary school, we were taught that scientists were geniuses who never made a mistake, and weren't we lucky to live in the era in which all truth had finally been discovered? And is it really much different today? What if we taught people not just what scientists had *found*, but the process of conjecture, failure, uncertainty, and testing by which they had found it? Of course scientists make mistakes, but what is special about them is that they embrace an ethos that champions turning to the evidence as a way to learn from them. What if we educated people about the values of science by demonstrating the importance of the scientist's creed: openness, humility, respect for uncertainty, honesty, transparency, and the courage to expose one's work to rigorous testing? I believe this kind of science education would do more to defeat science denial than anything else we could do.[66] It would teach children to think more like scientists and to channel what it means not to know something, then turn to the empirical evidence to find an answer; to work out the predictions of your model and then, if that prediction fails, live by the result. In this way we might give people an earlier understanding of the value of scientific uncertainty, and appreciate what we can learn from failure. Within this context, the facts of science might make more sense, and trust in scientists would grow accordingly. Anything that would encourage more people to identify with scientific values would be a step in the right direction.

As for the chasm between our political beliefs—will we ever find a way to cross it? What does teaching science have to do with politics? I submit that, just as post-truth started with science denial, perhaps a solution to the problem of science denial might help us to heal our post-truth politics. If people can learn to embrace scientific values, maybe they can change their values in other areas as well. To enlarge their circle of concern. To care more about the lives of people they may never meet.

For now, we have our hands full with the problem of science denial. So why not build an army of people to address that? There are a lot of people who believe in science, and a lot who care about climate change. And now—with Schmid and Betsch's results in hand—we understand that there is a way for *all* of us to make a difference. Walking away or refusing to engage with science deniers is the worst possible thing you can do. If your conversation partner is misinformed, continuing the conversation is the best way to try to get them to change their mind. So why not get out there and talk to science deniers? Try to change their identity and get them to think more like scientists? If I can go to a Flat Earth convention, you can talk to your niece or brother-in-law about anti-vaxx.

Of course, this is a lot of work, and it would be easier not to do it. How great would it be if people already thought like scientists, and all we had to do was give them the evidence! But that is not the reality we live in. Even people who trust science may not understand the thought process behind it. But that is the key to true understanding. And it is the key to helping them change their identity too. Many will take comfort in cynicism and the idea that in most instances this approach won't work. But that is entirely consistent with what Schmid and Betsch found. They did not promise that content rebuttal and technique rebuttal would always work. Instead, they gave us a way that would work if anything would. Yes, it is easier to treat those who deny science with contempt and refuse to engage with them. After all of my travels to engage with science deniers, I understand this frustration. But if we indulge it, we all lose.

Meanwhile, hard-core science deniers are out there right now spreading misinformation and recruiting new members. Those with selfish interests are all too happy to churn out disinformation and exploit any preexisting confusion or skepticism wherever they can. In the meantime, empirical truth is available for anyone who cares to find it—so why don't more people care to look? We can try to make them care more, but some just won't. They are happy to remain in a cocoon of partisanship, propaganda, and willful ignorance that tells them they already know the right answer. Which means that some of them will be prepared to deny the truth even when it is right in front of them. That is what is so frustrating in talking to a science denier. By the time you get to them, often it's not just their beliefs but their anti-science values that have begun to harden into cement. But of course they don't see it that way. No one self-identifies as a science denier. Often they see themselves as more scientific than the scientists. What you think of them, many will think of you. When you engage in conversation with a science denier, it's good to remember the rule that every novelist lives by: the villain is the hero of his own story.

In some ways, our job is both harder and easier than we might have imagined. The challenge is not just to get people to accept certain facts, or change their beliefs, but to begin to understand and appreciate how scientists have acquired their hard-won knowledge through a process of rigorous examination, cooperative testing, and tolerance of uncertainty, so that deniers might begin to identify more with the values (and reasoning processes) of scientists.

Of course, we should try to educate the next generation, but we are running out of time. Our kids will inherit the problem of climate change, and maybe they can solve it, but they aren't in charge yet. Too many of the people in charge today are science deniers, and they hold an unreasonable amount of power over our future. So the problem of how to deal with science deniers falls to all of us. Right now. You cannot change someone's beliefs against their will, nor can you usually get them to admit that there is something they don't already know. Harder still might be to get them to change their values or identity. But there is no easier path to take when dealing with science deniers. We must try to make them understand. We must try to get them care. But first we have to go out there, face-to-face, and begin to talk to them.

Epilogue

When I wrote this book, Donald Trump was still president and we were in the midst of a pandemic. While this book was going to press, Joe Biden was elected president of the United States and two coronavirus vaccines were developed and approved.

Will things get better now?

I hope so. president Biden has already put us back into the Paris Agreement and promised to reverse some of the Trump administration's most egregious rollbacks of auto emissions standards and other environmental regulations that have done so much to contribute to global warming. Beyond this, Biden has shown much more willingness than Trump to listen to scientists on a host of other topics, to include COVID-19.

But there is reason for caution.

We are still waiting for widespread distribution of a coronavirus vaccine, and it is not clear that, even once that is available, we will be able to convince everyone to take it. Conspiracy theories, and just plain deeply rooted skepticism, are still strong. Are the COVID vaccines safe? One hopes that the FDA and other government agencies would not have approved them unless they were. It's a matter of trusting not only science but also the guardians of scientific policy, some of whom have succumbed to political pressure under Trump. Whether liberal or conservative, one reality of the last four years is that much of this trust is gone.

Despite Trump's removal from office, Trumpism remains. The insurrrection at the US Capitol on January 6, 2021, showed just how deeply a "fact-free" ideology has penetrated this country and the terrible consequences that can follow. What ended in violence in Washington, DC, began at the Plaza Hotel in New York in 1953, where a handful of executives decided to "fight the science" by creating a disinformation campaign against facts that might hurt their business interests, perhaps unwittingly creating a blueprint for the future denial of *any* fact that did not fit someone's preferred reality. From scientific topics like climate change to political ones like election fraud, the problem of denialism now seems to have metastasized from the corporate to the ideological realm, from science to our larger culture. That said, there remain deeply entrenched interests around coal, oil, and other resources where some people stand to make a profit from the manufacture of doubt. And, of course, ignorance, cognitive bias, and the amplification of false ideas on social media are not going anywhere soon.

Science denial didn't start with Trump, though he certainly made it worse. It has been around at least since Galileo, and probably before that. But if science denial existed before Trump, it will likely exist after him too. Of course, that doesn't mean things can't get better. As we begin to talk to one another again, and attempt to heal the partisan divide that has done so much to polarize us not just on morality and values but even on empirical questions, I am hopeful that science will be part of that conversation. Though science denial has been around for a long time, it certainly got *worse* during an administration that questioned the facts about everything from the path of a hurricane, to whether California forest fires could be prevented by more raking, and whether bleach and light were potential treatments for the coronavirus and masks a good preventative measure during a pandemic.

But we should also remember that, as we saw in chapter 6, science denial is not an element of any one particular party or partisan point of view. The five tropes strategy is at work wherever people are motivated to believe something other than what the scientific consensus recommends. It is a long battle for science and reason. If we call it "won" and ignore it again, what will happen in the future? With our epistemic standards too, eternal vigilance is needed.

Remember also that science denial still exists all over the world. Even though Trump is out of office, the ideological rejection of science is entrenched in other countries, which have their own problems. Anti-vaxx is rampant in Italy. Coronavirus conspiracy theories led to the toppling of 5G cell towers in England. Flat Earthers make up a nonnegligible portion of the national population in Brazil. And remember the central role that Russia and China have played in whipping science denial by using their propaganda machines to spread disinformation, in search of more ideological division to undermine Western democracies.

So how will we learn to talk to one another again? Science denial is a lingering problem, both in the US and overseas. It will be a challenge to figure out a way to heal the wounds that have divided us for so long, and some of those are around scientific topics. What to do?

I still think the solution is to talk. The strategies for talking to science deniers recommended in this book are ones that we might use in general to get people to listen to reason, science, and logic on *many* areas where partisan belief runs counter to expert judgment. If we continue to demonize one another—or just ignore what a number of people have to say—we run the risk of polarization once again. Better, I think, is to try to bring science deniers back into the fold and show them how useful science can be. And how nice would it be for expertise to mean something again?

As we've seen, a good deal of science denial is motivated by fear, alienation, ideology, and identity. We can and should work on that, no matter who is in the White House. Complacency is the enemy. And science denial is not someone else's problem. In particular, we have no time to waste in trying to make up for all of the time we have lost on global warming. We can't just assume that everything will be all right, now that there has been a change in leadership in the US. As I have said throughout, climate denial is the most pressing and dangerous example of science denial in the world right now, and it will require all hands on deck to address it. Once COVID-19 has been defeated, the climate crisis will still be with us. And the clock is ticking.

Remember those Maldivian fishermen? Even if we now feel more confident that scientific facts and rational argument will play a more central role in public policy than it has in the recent past (so the information deficit might finally grow smaller), there is another deficit that looms as large as ever. How much do we care? Enough to cut consumption? To invest some serious money in alternative fuels? To change not just our beliefs but also our behavior?

In the process of learning how to talk to a science denier, we must take on the challenge of getting someone not just to change their beliefs but to enlarge the circle of concern that defines what they value. And we will succeed in that only by remembering our common humanity. To embrace the idea that someone who disagrees with us is still worth talking to is to make an investment in our fellow human beings and in our future together. While we are trying to get science deniers to enlarge their circle of concern, we must enlarge our own circle to include them. To have a difficult conversation with someone is to respect them enough to try to convince them that they are wrong. Rather than dismiss or avoid such encounters, I believe they are our best means of achieving renewed trust and empathy, which can bring about both epistemic and social change.

And that is a good thing, not just for trying to convince anti-vaxxers, anti-evolutionists, Flat Earthers, and climate change deniers that there is room for them on the team that celebrates science, but finally to conclude that—if we want to make the world a better place—everyone must count. Both the Maldivian fisherman and the Pennsylvania coal miner. Both the parent who is afraid of vaccines and the front-line health care worker during the COVID crisis. "No one cares," said that kid to me on the boat in the Maldives. I disagree. If we can build back some trust through renewed outreach, we just might be able to solve the problems of belief, caring, and action all at once.

The challenges we face are great, but the ingenuity of science is perhaps our greatest means of hope for the future. But the other is recognition of our common humanity. When it comes to the consequences from a warming planet or a killer pandemic, ultimately we are all on the same team.

Acknowledgments

As always, it is my privilege to publish this book with the MIT Press. I would like to thank the entire staff for their tremendous efforts on my behalf over the years. I am especially grateful to my editor, Phil Laughlin, for his trusted advice and guidance; to my production editor, Judith Feldmann; and to my copy editor, Rachel Fudge, who has rescued me from myself on a number of occasions.

For their contributions to the content of this book—by way of stimulating conversation and the example of their research—I would like to thank my friends and professional colleagues Quassim Cassam, Asheley Landrum, Stephan Lewandowsky, Michael Patrick Lynch, Richard Price, Derek Roff, Michael Shermer, and Bruce Sherwood. Although we had only one brief conversation, I would like to express special thanks to Cornelia Betsch and her colleague Philipp Schmid for their pathbreaking work that provides empirical evidence which shows that it *is* worthwhile to push back against science deniers.

In preparation for this book, I conducted a series of interviews with many folks whom I can name only by pseudonym. I thank them all. Alongside these I would like to thank the people I can name, which includes Linda Fox, Alex Mead, and Dave and Erin Ninehouser. For moral support in conducting this project, I couldn't have done better than to rely on the encouragement and advice of my friends Robyn Rosenfeld and Sam Shapson. I am uniquely grateful to Aaron Mertz of the Aspen Institute for his commitment to shining more light on the dangers of science denial. For giving me an intellectual home at the Center for Philosophy and History of Science at Boston University these past twenty years, I send my deepest thanks to Alisa Bokulich.

As usual, my philosophical pals Andy Norman and Jon Haber gave me terrific comments on various drafts of this manuscript, and much to think about in our conversations as it took shape. My friend Louis Kuchnir deserves special thanks for visiting me in Denver in November 2018, when I was attending the Flat Earth convention, and keeping me sane. Special thanks as well to my friend Laurie Prendergast for preparing the index.

To my wife, Josephine, who provided an island of calm and encouragement while I was writing this book—during what was easily the worst year of my life—I thank her for everything. Philosophy, logic, reason, and science have delineated the most important work of my life. But none of it compares to love.

Notes

Introduction

1. Lee McIntyre, *Post-Truth* (Cambridge, MA: MIT Press, 2018).

2. This story is well told in Naomi Oreskes and Erik Conway, *Merchants of Doubt: How a Handful of Scientists Obscured the Truth on Issues from Tobacco Smoke to Global Warming* (New York: Bloomsbury, 2010).

3. Tara Palmeri, "Trump Fumes over Inaugural Crowd Size," *Politico*, January 22, 2017, https://www.politico.com/story/2017/01/donald-trump-protesters-inauguration-233986.

4. IPCC, "Special Report: Global Warming of 1.5 degree C" (2018), https://www .ipcc.ch/sr15/.

5. Chris Mooney and Brady Dennis, "The World Has Just over a Decade to Get Climate Change under Control, UN Scientist Says," *Washington Post*, October 7, 2018, https://www.washingtonpost.com/energy-environment/2018/10/08/world-has-only -years-get-climate-change-under-control-un-scientists-say/; "Arctic Ice Could Be Gone by 2030," *Telegraph*, September 16, 2010, https://www.telegraph.co.uk/news/earth /earthnews/8005620/Arctic-ice-could-be-gone-by-2030.html; Coral Davenport, "Major Climate Report Describes a Strong Risk of Crisis as Early as 2040," *New York Times*, October 7, 2018, https://www.nytimes.com/2018/10/07/climate/ipcc-climate-report -2040.html; Mark Fischetti, "Sea Level Could Rise 5 Feet in New York City by 2100," *Scientific American*, June 1, 2013, https://www.scientificamerican.com/article/fischetti -sea-level-could-rise-five-feet-new-york-city-nyc-2100/; Mary Caperton Morton, "With Nowhere to Hide from Rising Seas, Boston Prepares for a Wetter Future," *Science News*, August 6, 2019, https://www.sciencenews.org/article/boston-adapting-rising-sea-level -coastal-flooding.

6. Somini Sengupta, "U.N. Chief Warns of a Dangerous Tipping Point on Climate Change," *New York Times*, September 10, 2018, https://www.nytimes.com/2018/09 /10/climate/united-nations-climate-change.html.

7. Lisa Friedman, "'I Don't Know That It's Man-Made,' Trump Says of Climate Change. It Is," *New York Times*, October 15, 2018, https://www.nytimes.com/2018/10/15/climate/trump-climate-change-fact-check.html.

8. Joe Keohane, "How Facts Backfire," *Boston.com*, July 11, 2010, http://archive.boston.com/news/science/articles/2010/07/11/how_facts_backfire/.

9. Julie Beck, "This Article Won't Change Your Mind," *Atlantic*, December 11, 2019, https://www.theatlantic.com/science/archive/2017/03/this-article-wont-change-your-mind/519093/; Elizabeth Kolbert, "Why Facts Don't Change Our Mind," *New Yorker*, February 27, 2017, https://www.newyorker.com/magazine/2017/02/27/why-facts-dont-change-our-minds.

10. Alexios Mantzarlis, "Fact-Checking Doesn't 'Backfire,' New Study Suggests," Poynter, November 2, 2016, https://www.poynter.org/fact-checking/2016/fact-checking-doesnt-backfire-new-study-suggests/.

11. Philipp Schmid and Cornelia Betsch, "Effective Strategies for Rebutting Science Denialism in Public Discussions," *Nature Human Behaviour* 3 (September 2019): 931–939, https://www.nature.com/articles/s41562-019-0632-4.epdf.

12. The Schmid and Betsch experiment will be explored in more detail in chapter 3. In the meantime here are links to a few popular media accounts of their results: Diana Kwon, "How to Debate a Science Denier," *Scientific American*, June 25, 2019, https://www.scientificamerican.com/article/how-to-debate-a-science-denier/; Laura Hazard Owen, "Yes, It's Worth Arguing with Science Deniers—and here Are Some Techniques You Can Use," Nieman Lab, June 28, 2019, https://www.niemanlab.org/2019/06/yes-its-worth-arguing-with-science-deniers-and-here-are-some-techniques-you-can-use/; Cathleen O'Grady, "Two Tactics Effectively Limit the Spread of Science Denialism," *Ars Technica*, June 27, 2019, https://arstechnica.com/science/2019/06/debunking-science-denialism-does-work-but-not-perfectly/; Susan Perry, "Science Deniers Can Be Effectively Rebutted, Study Finds," *MinnPost*, July 26, 2019, https://www.minnpost.com/second-opinion/2019/07/science-deniers-can-be-effectively-rebutted-study-finds/.

13. Arguably, they dealt only with the second. For details, see Sander van der Linden, "Countering Science Denial," *Nature Human Behaviour* 3 (June 24, 2019): 889–890, https://www.nature.com/articles/s41562-019-0631-5.

14. Michael Shermer, "How to Convince Someone When Facts Fail," *Scientific American*, January 1, 2017, https://www.scientificamerican.com/article/how-to-convince-someone-when-facts-fail/.

15. Lena H. Sun and Maureen O'Hagen, "'It Will Take Off Like Wildfire': The Unique Dangers of the Washington State Measles Outbreak," *Washington Post*, February 6, 2019, https://www.washingtonpost.com/national/health-science/it-will-take-off-like

-a-wildfire-the-unique-dangers-of-the-washington-state-measles-outbreak/2019/02
/06/cfd5088a-28fa-11e9-b011-d8500644dc98_story.html.

16. Rose Branigin, "I Used to Be Opposed to Vaccines. This Is How I Changed My
Mind," *Washington Post*, February 11, 2019, https://www.washingtonpost.com/opin
ions/i-used-to-be-opposed-to-vaccines-this-is-how-i-changed-my-mind/2019/02/11
/20fca654-2e24-11e9-86ab-5d02109aeb01_story.html.

17. Aris Folley, "NASA Chief Says He Changed Mind about Climate Change because
He 'Read a Lot,'" *The Hill*, June 6, 2018, https://thehill.com/blogs/blog-briefing-room
/news/391050-nasa-chief-on-changing-view-of-climate-change-i-heard-a-lot-of.

Chapter 1

1. Meghan Bartels, "Is the Earth Flat? Why Rapper B.o.B. and Other Celebrities Are
So Wrong," *Newsweek*, September 26, 2017, https://www.newsweek.com/bob-rapper
-flat-earth-earth-round-nasa-671140.

2. Though Irving has since recanted. Des Bieler, "Kyrie Irving Sorry for Saying Earth
Is Flat, Blames It on a YouTube 'Rabbit Hole,'" *Washington Post*, October 1, 2018,
https://www.washingtonpost.com/sports/2018/10/02/kyrie-irving-sorry-saying
-earth-is-flat-blames-it-youtube-rabbit-hole/.

3. According to a 2018 YouGov poll, 5 percent of Americans reported doubts about
the shape of the Earth, with 2 percent firmly believing it is flat. Hoang Nguyen,
"Most Flat Earthers Consider Themselves Very Religious," YouGov, April 2, 2018,
https://today.yougov.com/topics/philosophy/articles-reports/2018/04/02/most-flat
-earthers-consider-themselves-religious.

4. Though a majority of the speakers were white men.

5. I learned later that many at FEIC believe that the Flat Earth Society is a shill
group, created by those who wish to make the idea of Flat Earth look ridiculous.
Remember the Monty Python film *Life of Brian*, where there is a blood feud between
the "Judean People's Front" and the "People's Front of Judea," and you might get an
idea of how bitter this rivalry is.

6. I later found out that I wasn't fooling anybody. When we met again in the hallway, he asked, "Lee, why are you here?" Despite my vow to keep my cover for the
first twenty-four hours, I didn't want to lie, so I came clean and told him that I
wasn't a believer in Flat Earth but was a philosopher who was there to learn more
about their beliefs. He didn't seem upset by this and just went on to explain something about how flights could not be tracked south of the equator.

7. To my surprise, this included President Trump. All of the Flat Earthers I asked said
that they disliked him and thought he was in on the conspiracy merely because he was

a "world leader." Also, in one of the presentations someone had shown a photograph of Trump touching a glass globe, which proved that he was a "globalist." "What Was That Glowing Orb Trump Touched in Saudi Arabia?" *New York Times*, May 22, 2017, https://www.nytimes.com/2017/05/22/world/middleeast/trump-glowing-orb-saudi.html.

8. I did not bring up the fact that water too is subject to gravitational pull. What seemed most curious to me, however, was that it never seemed to occur to him to question whether Noah's flood had actually happened. That was a given, and I surmised that he was a Flat Earther at least in part because he was trying to reconcile his cosmological views with his religious ones. Of course, Newtonian physics is completely consistent with a planet whose entire surface could be covered with water, but I don't think he knew that.

9. This is to say that even though very few Christians believe in Flat Earth, almost all of the Flat Earthers I met (with a few notable exceptions) were fundamentalist Christians. While they did not seem to rely on their faith as scientific proof, they did seek empirical evidence that would make all of their beliefs—both spiritual and worldly—consistent. And it must be said that most Flat Earthers seemed to embrace their views with a fervor that was tantamount to religious conviction.

10. My favorite was the guy who had flown to Denver with a carpenter's level on the tray table in front of him. Since the bubble never moved, he took this as proof of Flat Earth.

11. They apparently also took it as proof that they were right when no one showed up to try to prove them wrong. There was a rumor at the conference that there was a meeting of physicists in Denver at the time, but then why hadn't any of them showed up to FEIC to refute them? If it was so easy, where were they? They must be scared because they knew that the Flat Earthers were right!

12. David Gee, "Almost All Flat Earthers Say YouTube Videos Convinced Them, Study Says," *Friendly Atheist*, February 20, 2019, https://friendlyatheist.patheos.com /2019/02/20/almost-all-flat-earthers-say-youtube-videos-convinced-them-study -says/ https://www.tandfonline.com/doi/full/10.1080/15213269.2019.1669461.

13. For a quick primer on basic Flat Earth beliefs, start here: Mark Sargent, "Flat Earth Clues Introduction," YouTube, February10, 2015, https://youtu.be/T8-YdgU-CF4.

14. Tom Coomes, "Mirage of Chicago Skyline Seen from Michigan Shoreline," *ABC 57*, April 29, 2015, https://www.abc57.com/news/mirage-of-chicago-skyline-seen -from-michigan-shoreline.

15. If you are interested in watching Skiba's presentation, I found it here: https://www.youtube.com/watch?v=oz35aaxJTik.

16. Indeed, sometimes the image can even appear upside down! Allison Eck, "The Perfectly Scientific Explanation for Why Chicago Appeared Upside Down in

Michigan," *PBS*, May 8, 2015, https://www.pbs.org/wgbh/nova/article/the-perfectly-scientific-explanation-for-why-chicago-appeared-upside-down-in-michigan/.

17. Alan Burdick, "Looking for Life on a Flat Earth," *New Yorker*, May 30, 2018, https://www.newyorker.com/science/elements/looking-for-life-on-a-flat-earth. Though note that the Flat Earthers have an account of how Eratosthenes's experiment might work for them as well.

18. Some of the phenomena the Flat Earthers took as evidence for their theory was easily answered by basic physics, but they still questioned it. If the Earth is round, why can you sometimes see the Sun and Moon in the sky at the same time? If a lunar eclipse is caused by the Earth's shadow, does this mean that the Earth has to be directly between the Sun and Moon? Ignorance is a poor foundation for doubt. Why didn't they just look it up?

19. Lee McIntyre, *The Scientific Attitude: Defending Science from Denial, Fraud, and Pseudoscience* (Cambridge, MA: MIT Press, 2019).

20. Why do you have to go up so high? Because the Earth is so large.

21. Alex Horton, "'Mad' Mike Huges, Who Wanted to Prove the Flat-Earth Theory, Dies in Homemade-Rocket Disaster," *Washington Post,* February 23, 2020, https://www.washingtonpost.com/science/2020/02/23/mad-mike-hughes-dead/.

22. Andrew Whalen, "'Behind the Curve' Ending: Flat Earthers Disprove Themselves with Own Experiments in Netflix Documentary," *Newsweek*, February 25, 2019, https://www.newsweek.com/behind-curve-netflix-ending-light-experiment-mark-sargent-documentary-movie-1343362.

23. If so, it would be the basis for a credible accusation of fraud. See chapter 7 of my book *The Scientific Attitude*.

24. This story unfortunately had a sad ending, for not only did the Flat Earther refuse to concede, he spent the next twenty years harassing Wallace. Esther Inglis-Arkell, "A Historic Experiment Shows Why We Might Not Want to Debate Fanatics," *Gizmodo*, August 27, 2014, https://io9.gizmodo.com/a-historic-experiment-shows-why-we-might-not-want-to-de-1627339811 For a similar, more contemporary story, see Jim Underdown, "The Salton Sea Flat Earth Test: When Skeptics Meet Deniers," *Skeptical Inquirer* 42, no. 6 (November/December 2018), https://skepticalinquirer.org/2018/11/the-salton-sea-flat-earth-test-when-skeptics-meet-deniers/.

25. For more on science deniers' mistaken conceptions of science, see chapter 2 of my book *The Scientific Attitude*.

26. This point is elegantly made by Nobel Prize–winning physicist Richard Feynman in one of his lectures, "The Essence of Science": https://www.youtube.com/watch?v=LIxvQMhttq4.

27. There is actually a hypothesis for the last one . . . a competitor to Flat Earthism that is perhaps even more ludicrous. Beckett Mufson, "Apparently, Some People Think the Earth Is Shaped Like a Donut," *Vice*, November 13, 2018, https://www.vice.com/en_us/article/mbyak8/apparently-some-people-believe-the-earth-is-shaped-like-a-donut-1.

28. Lee McIntyre, "The Price of Denialism," *New York Times*, November 7, 2015, https://opinionator.blogs.nytimes.com/2015/11/07/the-rules-of-denialism/. See also *The Scientific Attitude*, pp. 41–46.

29. Mick West, *Escaping the Rabbit Hole* (New York: Skyhorse Publishing, 2018).

30. One might think that they would also buy into all other forms of science denial, but for the Flat Earthers this was not true. While many were anti-vaxxers, virtually everyone I met was *not* a climate change denier. Since they believed that we were living in a domed enclosure (something like a terrarium), they were convinced that global warming was an urgent matter, although they did tend to believe that it was caused by the government messing with our weather rather than carbon pollution.

31. This conclusion has been borne out by empirical work as well. In her 2019 study, Asheley Landrum concluded that Flat Earthers are distinct in two ways: they have low scientific knowledge and a high disposition for conspiracy theories. Asheley Landrum, Alex Olshansky, and Othello Richards, "Differential Suspectibililty to Misleading Flat Earth Arguments on YouTube," *Media Psychology*, September 29, 2019, https://www.tandfonline.com/doi/full/10.1080/15213269.2019.1669461.

32. John Ingold, "We Went to a Flat-Earth Convention and Found a Lesson about the Future of Post-Truth Life," *Colorado Sun*, November 20, 2018, https://coloradosun.com/2018/11/20/flat-earth-convention-denver-post-truth/.

33. For an excellent psychological study of some of the motivations and causal influences in conversion to Flat Eartherism, see Alex Olshansky, Robert M. Peaslee, and Ashley Landrum, "Flat-Smacked! Converting to Flat-Eartherism," *Journal of Media and Religion*, July 2, 2020, https://www.tandfonline.com/doi/full/10.1080/15348423.2020.1774257?scroll=top&needAccess=true.

34. One of the best introductions to the concept of cognitive dissonance can be found in Leon Festinger's classic *When Prophecy Fails* (New York: Harper Torchbooks, 1964), which is about a 1950s UFO cult that believed the Earth was going to end on a specific date, so they waited on a mountaintop for a spaceship to pick them up. After the appointed time came and went, instead of giving up their belief they instead turned to the idea that the faith of their tiny group was so great that it had saved humanity.

35. Karl Popper, *The Logic of Scientific Discovery* (New York: Basic Books, 1959).

36. Since his name was not listed on the public program, I decided not to name him here.

37. Actually, the idea is not quite so far-fetched as it might seem. According to a 2019 *New York Times* article, NASA plans to open the International Space Station to commercial business, including tourism, in the next few years. Kenneth Chang, "Want to Buy a Ticket to the Space Station? NASA Says Soon You Can," *New York Times*, June 7, 2019, https://www.nytimes.com/2019/06/07/science/space-station-nasa.html. There are also private companies, such as Virgin Galactic, that have plans to offer "space tourism" flights. Michael Sheetz, "Virgin Galactic Flies Its First Astronauts to the Edge of Space, Taking One Step Closer to Space Tourism," *CNBC*, December 13, 2018, https://www.cnbc.com/2018/12/13/virgin-galactic-flight-could-send-first-astronauts-to-edge-of-space.html. The test flight went 51.4 miles into the air.

38. My idea here is that if you go out forty-five miles and the top of the Sears Tower disappears, but there is still a mirage on the horizon, presumably if you went out *far* enough, then the top of the Sears Tower *in the mirage* would disappear too. If it did, Flat Earth would be wrong.

39. Hugh Morris, "The Trouble with Flying to Antarctica—and the Airline That's Planning to Start," *Telegraph*, April 17, 2019, https://www.telegraph.co.uk/travel/travel-truths/do-planes-fly-over-antarctica/.

40. I no longer have the actual sheet of paper, but I think it was LATAM Flight 801. As I look back, however, I am not certain the route would have taken us directly over Antarctica. But no matter. Since the COVID-19 pandemic, while Australians have been banned from flying anywhere else in the world, Qantas Airlines is offering charter tourist flights over Antarctica, which go directly over the south magnetic pole. Allie Godfrey, "Antarctica Flights and Qantas Plan to Fly Travellers over the Frozen Continent from November," *7news Australia*, August 7, 2020, https://7news.com.au/news/travel/antarctica-flights-and-qantas-plan-to-fly-travellers-over-the-frozen-continent-from-november--c-1224156; see also: https://www.antarcticaflights.com.au/the-worlds-most-unique-scenic-flight.

41. See Nick Marshall, "The Longest Flight Time for a Commerical Airline, *USA Today*, last updated March 21, 2018, https://traveltips.usatoday.com/longest-flight-time-commercial-airline-109284.html; David Slotnick, "I Flew on Qantas' 'Project Sunrise,' a Nonstop Flight from New York to Sydney, Australia, That Took Almost 20 Hours and Covered Nearly 10,000 Miles—Here's What It Was Like," *Business Insider*, October 21, 2019, https://www.businessinsider.com/qantas-longest-flight-new-york-sydney-project-sunrise-review-pictures-2019-10#-and-a-light-monitor-23.

42. If it occurred today, there would be one more thing to say. One month after FEIC 2018, an adventurer named Colin O'Brady *walked* 921 miles across Antarctica unassisted, for the first time in human history. Adam Skolnick, "Colin O'Brady Completes Crossing of Antarctica with Final 32-Hour Push," *New York Times*, December 26, 2018, https://www.nytimes.com/2018/12/26/sports/antarctica-race-colin-obrady.html. Presumably, if he had a way to make sure that he was on a

straight-line course, this might count as proof for the Flat Earthers? Karen Gilchrist, "This 33-Year-Old Just Completed an Incredible World First. Here's How He Stayed Motivated along the Way," *CNBC*, December 14, 2018, https://www.cnbc.com/2018/12/14/how-to-stay-motivated-advice-from-colin-obrady-antarctic-crossing.html.

43. Looking back I should have said: "So that's it, then. There is no evidence that could convince you that your views are wrong. So I guess they're based on faith after all."

44. Because they were not named outside the program handed out at the event, I will not name them here.

45. This was a huge red flag, for it is a cult tactic.

46. Though it can be done. See Cara Westover, *Educated* (New York: Random House, 2018).

47. Sam Cowie, "Brazil's Flat Earthers to Get Their Day in the Sun," *Guardian*, November 6, 2019, https://www.theguardian.com/world/2019/nov/06/brazil-flat-earth-conference-terra-plana.

48. In a later email, he clarified that this was probably also due to the fact that if you look at a globe, it doesn't make much sense to fly straight over Antarctica, except for one possible route that may not be commercially viable.

49. Lee McIntyre, "The Earth Is Round," *Newsweek*, June 14, 2019, https://pocketmags.com/us/newsweek-europe-magazine/14th-june-2019/articles/590932/the-earth-is-round.

50. Lee McIntyre, "Call All Physicists," *American Journal of Physics* 87, no. 9 (September 2019), https://aapt.scitation.org/doi/pdf/10.1119/1.5117828.

51. Article: https://brucesherwood.net/?p=420. Model: tinyurl.com/FEmodel.

52. Will the Flat Earthers allow it? They should. In his speech, Robbie Davidson said that he'd love it if more physicists came to Flat Earth conventions. In an article in CNN, however, he was quoted as saying that they won't because they just say "you guys are dumb." Robert Pichetta, "The Flat-Earth Conspiracy Is Spreading Around the Globe. Does It Hide a Darker Core?" CNN, November 18, 2019, https://www.cnn.com/2019/11/16/us/flat-earth-conference-conspiracy-theories-scli-intl/index.html

Chapter 2

1. In his book *Reality Check: How Science Deniers Threaten Our Future* (Bloomington: Indiana University Press, 2013), Donald Prothero goes one step further and claims not only that there are "common threads" among the tactics used by science deniers, but that these were pioneered by Holocaust deniers (xv).

2. The Hoofnagle brothers originated a list of five common tropes of science denialism in a 2007 blog post, which has been studied and expanded by other researchers

(see https://scienceblogs.com/denialism/about); Pascal Diethelm and Martin McKee, "Denialism: What Is It and How Should Scientists Respond?" *European Journal of Public Health* 19, no. 1 (January 2009), https://academic.oup.com/eurpub/article /19/1/2/463780; John Cook, "5 Characteristics of Scientific Denialism," Skeptical Science, March 17, 2010, https://skepticalscience.com/5-characteristics-of-scientific -denialism.html; "A History of FLICC: The 5 Techniques of Science Denial," Skeptical Science, March 31, 2010, https://skepticalscience.com/history-FLICC-5 -techniques-science-denial.html; Stephan Lewandowsky, Michael E. Mann, Nicholas J. L. Brown, and Harris Friedman, "Science and the Public: Debate, Denial, and Skepticism," *Journal of Social and Political Psychology* 4, no. 2 (2016), https://jspp .psychopen.eu/article/view/604/html.

3. Quoted from Diethelm and McKee, "Denialism."

4. For more on the tricky concept of "warrant" in scientific reasoning, see pp. 41–46 of *The Scientific Attitude*, where I give an account of the technical reasons why science cannot rely on certainty, given Hume's problem of induction. I also discuss the important doctrine of fallibilism.

5. Daniel Kahneman, *Thinking Fast and Slow* (New York: Farrar, Straus & Giroux, 2011).

6. Naveena Sadasivam, "New Data Proves Cruz Wrong on Climate Change, Again," *Texas Observer*, January 22, 2016, https://www.texasobserver.org/new-temperature -data-proves-ted-cruz-is-still-wrong-about-climate-change/.

7. The classic account that science depends on such efforts at falsification is due to Karl Popper. I discuss Popper's work, and a number of the challenges it faces, on pp. 30–35 of *The Scientific Attitude*.

8. Based on my own experience, you would never make it past the first few items, as they would dismiss all scientific explanations or evidence that were inconsistent with their beliefs as biased, fake, or part of a conspiracy.

9. There are a number of outstanding works on the problems with conspiracy theories. An ideal place to start is Quassim Cassam's *Conspiracy Theories* (Cambridge: Polity Press, 2019). For a short, popular account of Cassam's work, see Quassim Cassam, "Why Conspiracy Theories Are Deeply Dangerous," *New Statesman*, October 7, 2019, https://www.newstatesman.com/world/north-america/2019/10/why-conspiracy -theories-are-deeply-dangerous. Other outstanding resources include Brian Keeley, "Of Conspiracy Theories," *Journal of Philosophy* 96, no. 3 (March 1999): 109–126, and the following books: Mick West, *Escaping the Rabbit Hole* (New York: Skyhorse, 2018); Michael Shermer, *The Believing Brain* (New York: Holt, 2011); Donald Prothero, *Reality Check* (Bloomington: Indiana University Press, 2013); and Sara and Jack Gorman, *Denying to the Grave* (New York: Oxford University Press, 2017).

10. For a discussion of the distinction between "real" conspiracies and conspiracy theories, see West, *Escaping the Rabbit Hole*, xii. See also Stephan Lewandowsky and John Cook,

The Conspiracy Theory Handbook (2020), https://www.climatechangecommunication
.org/conspiracy-theory-handbook/.

11. West discusses this problem of "false" conspiracy theories throughout his book.

12. Eric Oliver and Thomas Wood give this definition in a *Washington Post* interview: John Sides, "Fifty Percent of Americans Believe in Some Conspiracy Theory.
Here's Why," *Washington Post*, February 19, 2015, https://www.washingtonpost.com
/news/monkey-cage/wp/2015/02/19/fifty-percent-of-americans-believe-in-some
-conspiracy-theory-heres-why/.

13. Cassam, "Why Conspiracy Theories Are Deeply Dangerous."

14. "But they could be true!" remarks the conspiracy theorist. But that isn't the
point. Without evidence, how do you judge the probability that any given belief
is more likely to be true than any another? Or between credulity and satire? Yes,
perhaps it is true that all birds died over fifty years ago and have been replaced with
cleverly disguised surveillance drones that were created by the government to spy
on us, but where is the evidence for that? Fernando Alfonso III, "Are Birds Actually
Government-Issued Drones? So Says a New Conspiracy Theory Making Waves (and
Money)," *Audubon*, November 16, 2018, https://www.audubon.org/news/are-birds
-actually-government-issued-drones-so-says-new-conspiracy-theory-making. Here we
see that conspiracy theorists exploit the inherent uncertainty of science to try to
make their own claims seem more credible. (See trope 5 ahead for more on this.)

15. J. Eric Oliver and Thomas J. Wood, "Conspiracy Theories and the Paranoid
Style(s) of Mass Opinion," *American Journal of Political Science*, March 5, 2014, https://
onlinelibrary.wiley.com/doi/abs/10.1111/ajps.12084.

16. For a popular account of Oliver and Wood's findings, see Sides, "Fifty-Percent of
Americans Believe in Some Conspiracy Theories."

17. Kim Komando, "The Great 5G Coronavirus Conspiracy," *USA Today*, April 20,
2020, https://www.usatoday.com/story/tech/columnist/2020/04/20/dispelling-belief
-5-g-networks-spreading-coronavirus/5148961002/. If you think these are bad, consider the one about how the British royal family are actually reptiles. "The Reptilian
Elite," *Time*, http://content.time.com/time/specials/packages/article/0,28804,1860871
_1860876_1861029,00.html.

18. Jan-Willem van Prooijen and Karen M. Douglas, "Conspiracy Theories as Part
of History: The Role of Societal Crisis Situations," *Memory Studies*, June 29, 2017,
https://journals.sagepub.com/doi/10.1177/1750698017701615.

19. Jeremy Schulman, "Every Insane Thing Donald Trump Has Said about Global
Warming," *Mother Jones*, December 12, 2018, https://www.motherjones.com/environ
ment/2016/12/trump-climate-timeline/.

20. Does this mean that there is a causal link between conspiracy theories and
science denial? According to at least one researcher, yes. As reported by the BBC

in 2018, Stephan Lewandowsky found that "the stronger a person believes in a conspiracy, the less likely they are to trust scientific facts. It is more likely they will think the person attempting to reason with them is in on it." Melissa Hoogenboom, "The Enduring Appeal of Conspiracy Theories," *BBC*, January 24, 2018, https://www.bbc.com/future/article/20180124-the-enduring-appeal-of-conspiracy -theories.

21. An important related question is why anyone would invent or peddle a conspiracy theory, whether they believed it or not. Quassim Cassam has argued that "conspiracy theories are first and foremost forms of political propaganda." See Cassam, "Why Conspiracy Theories Are Deeply Dangerous." For more, see his fascinating book, *Conspiracy Theories* (Cambridge: Polity, 2019).

22. Aleksandra Cichocka, Marta Marchlewska, and Golec de Zavala, "Does Self-Love or Self-Hate Predict Conspiracy Beliefs? Narcissism, Self-Esteem, and the Endorsement of Conspiracy Theories," *Social Psychological and Personality Science*, November 13, 2015, https://journals.sagepub.com/doi/abs/10.1177/1948550615616170; Joseph Vitriol and Jessecae K. Marsh, "The Illusion of Explanatory Depth and Endorsement of Conspiracy Beliefs," *European Journal of Social Psychology*, May 12, 2018, https:// onlinelibrary.wiley.com/doi/abs/10.1002/ejsp.2504; Christopher M. Federico, Allison L. Williams, and Joseph A. Vitriol, "The Role of System Identity Threat in Conspiracy Theory Endorsement," *European Journal of Social Psychology*, April 18, 2018, https:// onlinelibrary.wiley.com/doi/abs/10.1002/ejsp.2495.

23. One outstanding and handy resource for understanding what causes conspiracy theories, and how to deal with them, is Lewandowsky and Cook's *The Conspiracy Theory Handbook*.

24. Anthony Lantian, Dominique Muller, Cecile Nurra, and Karen M. Douglas, "'I Know Things They Don't Know': The Role of Need for Uniqueness in Belief in Conspiracy Theories," *Social Psychology*, July 10, 2017, https://econtent.hogrefe.com/doi /10.1027/1864-9335/a000306.

25. Roland Imhoff and Pia Karoline Lamberty, "Too Special to Be Duped: Need for Uniqueness Motivates Conspiracy Beliefs," *European Journal of Social Psychology*, May 23, 2017, https://onlinelibrary.wiley.com/doi/abs/10.1002/ejsp.2265; Roland Imhoff, "How to Think Like a Conspiracy Theorist," *Aeon*, May 5, 2018, https:// theweek.com/articles/769349/how-think-like-conspiracy-theorist; Jon Stock, "Why We Can Believe in Almost Anything in This Age of Paranoia," *Telegraph*, June 4, 2018, https://www.telegraph.co.uk/property/smart-living/age-of-paranoia/.

26. See Oreskes and Conway, *Merchants of Doubt*.

27. Tom Nichols, *The Death of Expertise: The Campaign against Established Knowledge and Why It Matters* (Oxford: Oxford University Press, 2017).

28. This, by the way, is why science denial is selective. Where an empirical matter does not tread on ideological turf, who cares? Flat Earthers fly on planes and use cell

phones, because those technologies do not clash with their preferred beliefs. But when they do, all of a sudden scientists become evil.

29. D. Piepgrass, "Climate Science Denial Explained: Tactics of Denial," Skeptical Science, April 17, 2018, https://skepticalscience.com/agw-denial-explained-2.html; https://skepticalscience.com/graphics.php?g=227.

30. John Cook, "The 5 Characteristics of Science Denialism," Skeptical Science, March 17, 2010, https://skepticalscience.com/5-characteristics-of-scientific -denialism.html.

31. For an excellent discussion of the fallacies of informal logic, see Douglas Walton, *Informal Logic* (Cambridge: Cambridge University Press, 1989).

32. For an excellent resource on climate denial, see Piepgrass, "Climate Science Denial Explained," https://skepticalscience.com/agw-denial-explained.html.

33. For a rigorous but accessible account of these ideas, see Hugh Mellor's out-standing lecture, "The Warrant of Induction," https://www.repository.cam.ac.uk /bitstream/handle/1810/3475/InauguralText.html?sequence=5&isAllowed=y.

34. For that matter, why did so many of them trust the planes they flew in on, believing that their pilots were part of a worldwide conspiracy that is headed by the devil?

35. Theodosius Dobzhansky, "Nothing in Biology Makes Sense Except in Light of Evolution," *American Biology Teacher*, March 1973, https://www.pbs.org/wgbh /evolution/library/10/2/text_pop/l_102_01.html.

36. Yes, they did, but he convinced them with evidence. See McIntyre, *The Scientific Attitude*, 65.

37. Alister Doyle, "Evidence for Man-Made Global Warming Hits 'Gold Standard': Scientists," *Reuters*, February 25, 2019, https://www.reuters.com/article/us-climatechange -temperatures/evidence-for-man-made-global-warming-hits-gold-standard-scientists -idUSKCN1QE1ZU.

38. Note here again the double standard. For a disfavored belief the conspiracy theorist will say, "You can't prove it's true," but for a favored one they will say, "You can't prove it's *not* true."

39. See Stephan Lewandowsky and Karl Oberauer, "Motivated Rejection of Science," *Current Directions in Psychological Science* 25, no. 4 (2016): 217–222; and Brendan Nyhan and Jason Reifler, "When Corrections Fail," *Political Behavior* 32 (2010): 303–330.

40. One might wonder, though, whether this means that *all* science denial has to be cynically manufactured. If so, what is the special group that is benefiting from Flat Earth?

41. A detailed history of these events can be found in Oreskes and Conway, *Merchants of Doubt*. See also Mike Stobbe, "Historic Smoking Report Marks 50th Anniversary," *USA Today*, January 5, 2014, https://www.usatoday.com/story/money/business/2014/01/05/historic-smoking-report-marks-50th-anniversary/4318233/

42. Oreskes and Conway, *Merchants of Doubt*.

43. Graham Readfern, "Doubt over Climate Science Is a Product with an Industry Behind It," *Guardian*, March 5, 2015, https://www.theguardian.com/environment/planet-oz/2015/mar/05/doubt-over-climate-science-is-a-product-with-an-industry-behind-it.

44. Oreskes and Conway, *Merchants of Doubt*, 34–35.

45. It came out later, during litigation in the 1990s, that the tobacco companies knew all along exactly how harmful their products were. Incredibly, the same thing happened to big oil forty years later, when they used the same strategy—and in some cases even the same researchers—to create doubt about climate change. Benjamin Hulac, "Tobacco and Oil Industries Used Same Researchers to Sway Public," *Scientific American*, July 20, 2016, https://www.scientificamerican.com/article/tobacco-and-oil-industries-used-same-researchers-to-sway-public1/.

46. Shannon Hall, "Exxon Knew about Climate Change Almost 40 Years Ago," *Scientific American*, October 26, 2015, https://www.scientificamerican.com/article/exxon-knew-about-climate-change-almost-40-years-ago/.

47. Remember that a liar can sometimes come to believe their own lie. In *The Folly of Fools (New York: Basic Books, 2011)*, Robert Trivers goes into some detail about the cognitive and psychological processes that can lead one to delusion. Repetition of a lie is one means. We've all met the person who lies so much that they begin to believe it. And, indeed, this can lead them down a slippery slope whereby even someone who starts in ignorance (or as a liar) can become willfully ignorant, and end up in full-blown denial. Compare my discussion of this phenomenon in *Respecting Truth*, 79–80.

48. Richard Nisbett and Timothy Wilson, "Telling More Than We Can Know," *Psychological Review* 84, no. 3 (1977), http://people.virginia.edu/~tdw/nisbett&wilson.pdf.

49. See Trivers, *The Folly of Fools*. See also Keith Kahn-Harris, *Denial: The Unspeakable Truth* (London: Notting Hill Editions, 2018).

50. Darren Schrieber et al., "Red Brain, Blue Brain: Evaluative Processes Differ in Democrats and Republicans," *PLoS One*, February 13, 2013, https://www.ncbi.nlm.nih.gov/pmc/articles/PMC3572122/. See also the work of Jonas Kaplan, who hooked partisans up to an fMRI machine and had them read opinions that challenged their beliefs. He measured greater blood flow to the part of the brain associated with basic

beliefs and personal identity. "The Partisan Brain," *Economist*, December 8, 2018, https://www.economist.com/united-states/2018/12/08/what-psychology-experiments -tell-you-about-why-people-deny-facts; Jonas T. Kaplan, Sarah I. Gimbel, and Sam Harris, "Neural Correlates of Maintaining One's Political Beliefs in the Face of Coun- terevidence," *Scientific Reports*, December 23, 2016, https://www.ncbi.nlm.nih.gov /pmc/articles/PMC5180221/.

51. John Ridgway, "The Neurobiology of Climate Denial," *The Global Warming Policy Forum*, August 6, 2018, https://www.thegwpf.com/the-neurobiology-of-climate -change-denial/.

52. Anna Merlan, "Everything I Learned While Getting Kicked Out of America's Big- gest Anti-Vaccine Conference," *Jezebel*, June 20, 2019, https://jezebel.com/everything -i-learned-while-getting-kicked-out-of-americ-1834992879.

53. For this reason, some recommend not using the term "science denier." Or instead calling anti-vaxxers "vaccine-hesitant." I defend my usage of the stronger terms when discussing the problem with fellow researchers, but perhaps not when face-to-face with a science denier.

54. One quite interesting psychological analysis of the conversion process to Flat Eartherism—including a few remarks about identity—is Alex Olshansky, Robert M. Peaslee, and Asheley R. Landrum, "Flat-Smacked! Converting to Flat Eartherism," *Journal of Media & Religion* 19, no. 2 (in press), 46–59, doi:10.1080/15348423.2020. 1774257.

55. See McIntyre, *Post-Truth*, chapter 2.

56. In a 2019 Pew survey, 96 percent of liberal Democrats said that human activity had at least some effect on climate change, as opposed to only 53 percent of con- servative Republicans. Cary Funk and Meg Hefferson, "U.S. Public Views on Climate and Energy," Pew Research Center, November 25, 2019, https://www.pewresearch .org/science/2019/11/25/u-s-public-views-on-climate-and-energy/.

57. Lewandowsky and Oberauer, "Motivated Rejection of Science," 2016.

58. See Kahneman, *Thinking Fast and Slow*.

59. Michael Lynch, *Know-It-All Society: Truth and Arrogance in Political Culture* (New York: Liveright, 2019), 6.

60. Dan Kahan et al., "Motivated Numeracy and Enlightened Self-Government," *Behavioural Public Policy* (preprint, 2013), https://pdfs.semanticscholar.org/2125/a9 ade77f4d1143c4f5b15a534386e72e3aea.pdf.

61. "The Case for a 'Deficit Model' of Science Communication," *Sci Dev Net*, June 27, 2005, https://bit.ly/2AQ7mT1.

62. Kahan et al., "Motivated Numeracy and Enlightened Self-Government."

63. Ezra Klein, "How Politics Makes Us Stupid," *Vox*, April 6, 2014, https://www.vox.com/2014/4/6/5556462/brain-dead-how-politics-makes-us-stupid.

64. Kahan et al., "Motivated Numeracy."

65. Klein, "How Politics Makes Us Stupid."

66. Lilliana Mason, "Ideologues without Issues: The Polarizing Consequences of Ideological Identities," *Public Opinion Quarterly*, March 21, 2018, https://academic.oup.com/poq/article/82/S1/866/4951269.

67. When asked, "How would you feel about living next door to a liberal?," the reaction was much worse for conservatives than when they were asked, "How would you feel about living next door to someone who supported abortion rights for women?" The same was true for how liberals felt about conservatives.

68. Tom Jacobs, "Ideology Isn't Really about Issues," *Pacific Standard*, April 30, 2018, https://psmag.com/news/turns-out-its-all-identity-politics; Zaid Jilani, "A New Study Shows How American Polarization Is Driven by Team Sports Mentality, Not by Disagreement on Issues," *Intercept*, April 3, 2018, https://theintercept.com/2018/04/03/politics-liberal-democrat-conservative-republican/; Cameron Brick and Sander van der Linden, "How Identity, Not Issues, Explains the Partisan Divide," *Scientific American*, June 19, 2018, https://www.scientificamerican.com/article/how-identity-not-issues-explains-the-partisan-divide/.

69. Kwame Anthony Appiah, "People Don't Vote for What They Want. They Vote for Who They Are," *Washington Post*, August 30, 2018, https://www.washingtonpost.com/outlook/people-dont-vote-for-want-they-want-they-vote-for-who-they-are/2018/08/30/fb5b7e44-abd7-11e8-8a0c-70b618c98d3c_story.html.

70. And how difficult to get them to change back—merely on the basis of scientific evidence—without considering the effect this might have on their identity?

71. Solomon Asch, "Opinions and Social Pressure," *Scientific American*, November 1955, https://www.lucs.lu.se/wp-content/uploads/2015/02/Asch-1955-Opinions-and-Social-Pressure.pdf.

72. Peter Boghossian and James Lindsay, *How to Have Impossible Conversations* (New York: Lifelong Books, 2019), 99–100.

73. Boghossian and Lindsay, *How to Have Impossible Conversations*, 103.

74. See Karl Popper, *Conjectures and Refutations* (New York: Harper Torchbooks, 1965), ch. 1.

75. As Lynch puts it, "Attacks on our convictions seem like attacks on our identity—because they are" (*Know-It-All Society*, 6).

76. Dan Kahan, "What People 'Believe' About Global Warming Doesn't Reflect What They Know; It Expresses Who They Are," The Cultural Cognition Project at

Yale Law School, April 23, 2014, http://www.culturalcognition.net/blog/2014/4/23
/what-you-believe-about-climate-change-doesnt-reflect-what-yo.html.

77. This is why science denial is more than just the rejection of particular scientific
facts or consensus. It is more fundamentally a rejection of the scientists' creed: that
empirical questions should be decided on the basis of evidence. Ideology, or one's
personal desire for a theory to be true, should have nothing to do with it, because
scientific consensus is built not on what scientists want to believe but on what they
are compelled to believe by the process of rigorous testing and analysis that drives
them to the same conclusion. See McIntyre, *The Scientific Attitude*, 47–52.

Chapter 3

1. James H. Kuklinski et al., "Misinformation and the Currency of Democratic Citizenship," *Journal of Politics* 62, no. 3 (August 2000), https://www.uvm.edu/~dguber
/POLS234/articles/kuklinski.pdf.

2. This is called the Dunning-Kruger effect. I discuss it in *Post-Truth*, 51–58.

3. A telephone survey is surely not as personal as a face-to-face encounter, but it is
more personal than interacting with subjects only online. Kuklinski's work involved
a half-hour phone conversation with each of his participants. That is a long time to
be on the phone with someone.

4. Although it did not explicitly address the issue of partisan identity, note the care
with which the Kuklinski study dealt with the potential problem of cognitive dissonance and polarization. They took care *not* to tweak the ego of the subjects by
shoving it in their faces that they had just changed their minds. Instead of saying,
"Well, just a minute ago I would've thought you were totally against welfare," they
rephrased the question from one about what they thought the ideal level of welfare
should be to whether they supported welfare payments.

5. Kuklinski et al., "Misinformation and the Currency of Democratic Citizenship."

6. David Redlawsk et al., "The Affective Tipping Point: Do Motivated Reasoners Ever
'Get It'?," *Political Psychology*, July 12, 2010, https://onlinelibrary.wiley.com/doi/10
.1111/j.1467-9221.2010.00772.x.

7. Indeed, researchers noted that subjects react initially to how the information
makes them feel, second to its content. Of course, we also react emotionally when a
belief is in our favor. Note the well known problem of "confirmation bias."

8. See Michael Lynch, *Know-It-All Society* (New York: Liveright, 2019).

9. Redlawsk et al., "The Affective Tipping Point," 589. One might consider that
there is a social factor here too, over whether subjects were embarrassed to stick
with their original choice. If social factors like peer pressure or identity can influence
belief formation surely they can influence belief change as well.

10. Redlawsk et al., "The Affective Tipping Point," 590.

11. Brendan Nyhan and Jason Reifler, "When Corrrections Fail: The Persistence of Political Misperceptions," *Political Behavior (preprint)*, 2010, https://www.dartmouth .edu/~nyhan/nyhan-reifler.pdf.

12. Julie Beck, "This Article Won't Change Your Mind," *Atlantic*, March 13, 2017, https://www.theatlantic.com/science/archive/2017/03/this-article-wont-change -your-mind/519093/; Elizabeth Kolbert, "Why Facts Don't Change Our Minds," *New Yorker*, February 27, 2017, https://www.newyorker.com/magazine/2017/02/27/why -facts-dont-change-our-minds.

13. Thomas Wood and Ethan Porter, "The Elusive Backfire Effect: Mass Attitudes' Steadfast Factual Adherence," *Political Behavior*, January 6, 2018, http://dx.doi.org /10.2139/ssrn.2819073.

14. Eileen Dombrowski, "Facts Matter After All: Rejecting the Backfire Effect," Oxford Education Blog, March 12, 2018, https://educationblog.oup.com/theory-of -knowledge/facts-matter-after-all-rejecting-the-backfire-effect.

15. Ethan Porter and Thomas Wood, "No, We're Not Living in a Post-Fact World," *Politico*, January 4, 2020, https://www.politico.com/news/magazine/2020/01/04 /some-good-news-for-2020-facts-still-matter-092771; Alexios Mantzarlis, "Fact Checking Doesn't 'Backfire,'" Poynter, November 2, 2016, https://www.poynter .org/fact-checking/2016/fact-checking-doesnt-backfire-new-study-suggests/.

16. Brendan Nyhan and Jason Reifler, "The Role of Information Deficits and Identity Threat in the Prevalence of Misperceptions," *Journal of Elections, Public Opinions and Parties*, May 6, 2018, https://www.dartmouth.edu/~nyhan/opening-political -mind.pdf.

17. Quoted from the abstract of Nyhan and Reifler, "The Role of Information Deficits and Identity Threat."

18. See Lynch, *Know-It-All Society*.

19. Michael Shermer, "How to Convince Someone When Facts Fail," *Scientific American*, January 1, 2017, https://www.scientificamerican.com/article/how-to-convince -someone-when-facts-fail/.

20. Recall Leon Festinger's classic work *When Prophecy Fails* (New York: Harper, 1964)—already discussed in note 34 of chapter 1—where he tells the story of a doomsday cult called The Seekers who gave away all of their worldly possessions and sat on a mountaintop waiting to be rescued by an alien spaceship before the impending apocalypse. When nothing happened, they did not change their beliefs but instead embraced the idea that it was their faith that had saved the world.

21. See the quotation accompanying note 14 in the introduction. Shermer, "How to Convince Someone When Facts Fail."

22. Philipp Schmid and Cornelia Betsch, "Effective Strategies for Rebutting Science Denialism in Public Discussions," *Nature Human Behaviour*, June 24, 2019, https://www.nature.com/articles/s41562-019-0632-4.

23. Cathleen O'Grady, "Two Tactics Effectively Limit the Spread of Science Denialism," *Ars Technica*, June 27, 2019, https://arstechnica.com/science/2019/06/debunking-science-denialism-does-work-but-not-perfectly/.

24. O'Grady, "Two Tactics."

25. Diana Kwon, "How to Debate a Science Denier," *Scientific American*, June 25, 2019, https://www.scientificamerican.com/article/how-to-debate-a-science-denier/.

26. Schmid and Betsch, "Effective Strategies," 5.

27. Apparently confirmation bias is a powerful force not just for science deniers but for those who study them as well.

28. One problem is that many scientists have never been educated about how to share their research with a lay audience. The Alan Alda Center at Stonybrook is trying to help scientists and others learn more about the public communication of science (https://www.aldacenter.org). Of course, scientists sometimes protest and do other public events where they stand up for science, like the 2017 March for Science. But some have argued that this is a bad idea because it can be polarizing and makes scientists seem like just another special interest group. Instead, one commentator has said, "Rather than marching on Washington and in other locations around the country, I suggest that my fellow scientists march in local civic groups, churches, schools, county fairs and, privately, into the offices of elected officials. make contact with that part of America that doesn't know any scientists. Put a face on the debate. Help them understand what we do, and how we do it. Give them your email, or better yet, your phone number." Robert S. Young, "A Scientists' March on Washington Is a Bad Idea," *New York Times,* January 31, 2017, https://www.nytimes.com/2017/01/31/opinion/a-scientists-march-on-washington-is-a-bad-idea.html.

29. Young, "A Scientists' March on Washington."

30. Schmid and Betsch do not compare the effectiveness of "technique rebuttal" (after the fact) with "inoculation" (before the fact) through exposure to the sketchy rhetorical devices used by science deniers. In earlier work, Sander van der Linden discusses additional evidence for the effectiveness of inoculation theory. Sander van der Linden, "Inoculating against Misinformation," *Science*, December 1, 2017, https://science.sciencemag.org/content/358/6367/1141.2. In even more recent work, he has invented an online "fake news game" that has had some success in "pre-bunking" against scientific disinformation. Sander van der Linden, "Fake News 'Vaccine' Works: 'Pre-bunking Game Reduces Susceptibility to Disinformation," *Science Daily*, June 24, 2019, https://www.sciencedaily.com/releases/2019/06/190624204800.htm.

31. This raises a potentially important criticism of Schmid and Betsch's study—that was explored by Sander van der Linden in a discussion of their study in the same issue of *Nature Human Behaviour*—which is that their approach is entirely reactive. Van der Linden explores an approach called "pre-bunking," where he tries to "inoculate" the subject against scientific misinformation before they have even heard it. Is this more effective than the kind of debunking that Schmid and Betsch engage in? No one knows. Van der Linden's work is superb. Sander van der Linden, "Countering Science Denial," *Nature Human Behaviour* 3 (2019): 889–890, https://www.nature .com/articles/s41562-019-0631-5. His "Bad News" game helps subjects to see the mistakes of the five common errors script coming before they are even exposed to them. Sander van der Linden, "Bad News: A Psychological 'Vaccine' against Fake News," *Informm*, September 7, 2019, https://informm.org/2019/09/07/bad-news-a -psychological-vaccine-against-fake-news-sander-van-der-linden-and-jon-rozenbeek/. Other excellent work on the idea of inoculation against scientific misinformation can be found in John Cook, Stephan Lewandowsky, and Ulrich Ecker, "Neutralizing Misinformation through Inoculation," *PLoS One*, May 5, 2017, https://journals.plos .org/plosone/article?id=10.1371/journal.pone.0175799.

32. Note that even if it were, however, it still might be important to weigh in. Every lie has an audience, and if we do not confront the science deniers when they are recruiting new members, their misinformation will spread.

33. In *The Scientific Attitude*, I develop a theory that the most essential thing about science is its values. Might teaching these values help to convert science deniers?

34. Quoted from Adrian Bardon, *The Truth about Denial* (Oxford: Oxford University Press, 2020), 86.

35. In *The Truth about Denial*, Adrian Bardon discusses Heather Douglas's idea that we need to educate the public not just on scientific facts but on how science actually works. It is a rigorous process, with a unique culture and set of values. If we shared this, might it increase people's identification with scientists (see Bardon, *The Truth about Denial,* 300)? Sara Gorman and Jack Gorman, in *Denying to the Grave* (Oxford: Oxford University Press, 2017, 22), make a largely similar point about how we educate children.

36. Mick West, *Escaping the Rabbit Hole: How to Debunk Conspiracy Theories Using Facts, Logic, and Respect* (New York: Skyhorse, 2018), 60.

37. Peter Boghossian and James Lindsay, *How to Have Impossible Conversations* (New York: Lifelong Books, 2019), 50–51.

38. Boghossian and Lidsay, *How to Have Impossible Conversations*, 12.

39. The closest I have been able to find is John Cook and Stephan Lewandowky, who provide advice for dealing with conspiracy theorists in *The Conspiracy Theory Handbook*, https://www.climatechangecommunication.org/conspiracy-theory-handbook/.

Recall Michael Shermer's general advice drawn from his previously cited article in *Scientific American*, "How to Convince Someone When Facts Fail," https://www .scientificamerican.com/article/how-to-convince-someone-when-facts-fail/.

40. Why not? This seems like ripe ground for further research.

41. Jonathan Berman, *Anti-Vaxxers* (Cambridge, MA: MIT Press, 2020). See also Seth Mnookin, *The Panic Virus* (New York: Simon and Schuster, 2011).

42. *The Scientific Attitude*, 143–147; *Respecting Truth*, 46–47; "Could a Booster Shot of Truth Help Scientists Fight the Anti-vaccine Crisis?" *The Conversation*, March 8, 2019, https://theconversation.com/could-a-booster-shot-of-truth-help-scientists -fight-the-anti-vaccine-crisis-111154; "Public Belief Formation and the Politiciza-tion of Vaccine Science," *The Critique*, September 10, 2015, http://www.thecritique .com/articles/public-belief-formation-the-politicization-of-vaccine-science-a-case -study-in-respecting-truth/.

43. See *The Scientific Attitude*, 143–147. It is important to realize that the anti-vaxx phenomenon goes back to the very first vaccine, but after Wakefield's work it became much more widespread. For a complete history, see Berman, *Anti-vaxxers*.

44. Associated Press, "Clark County Keeps 800 Students Out of School Due to Measles Outbreak," *NBC News*, March 7, 2019, https://www.nbcnews.com/storyline /measles-outbreak/clark-county-keeps-800-students-out-school-due-measles-outbreak -n980491.

45. Lena Sun and Maureen O'Hagen, "'It Will Take Off Like Wildfire': The Unique Dangers of the Washington State Measles Outbreak," *Washington Post*, February 6, 2019, https://www.washingtonpost.com/national/health-science/it-will-take-off -like-a-wildfire-the-unique-dangers-of-the-washington-state-measles-outbreak /2019/02/06/cfd5088a-28fa-11e9-b011-d8500644dc98_story.html?utm_term=.5b659 64ef193.

46. Rose Branigin, "I Used to Be Opposed to Vaccines," *Washington Post*, Febru-ary 11, 2019, https://www.washingtonpost.com/opinions/i-used-to-be-opposed-to -vaccines-this-is-how-i-changed-my-mind/2019/02/11/20fca654-2e24-11e9-86ab -5d02109aeb01_story.html?utm_term=.089a62aac347.

47. Vanessa Milne et al., "Seven Ways to Talk to Anti-vaxxers (That Actually Might Change Their Minds)," *Healthy Debate*, August 31, 2017, https://healthydebate.ca /2017/08/topic/vaccine-safety-hesitancy. In *Anti-vaxxers*, Jonathan Berman tells the story of seven or eight other anti-vaxxers who changed their minds under similar circumstances (205–209).

48. Marina Koren, "Trump's NASA Chief: 'I Fully Believe and Know the Climate Is Changing,'" *Atlantic*, May 17, 2018, https://www.theatlantic.com/science/archive/2018 /05/trump-nasa-climate-change-bridenstine/560642/.

49. Terry Gross, "How a Rising Star of White Nationalism Broke Free from the Movement," *NPR*, September 24, 2018, https://www.npr.org/2018/09/24/651052970/how -a-rising-star-of-white-nationalism-broke-free-from-the-movement.

50. Other conversions along these lines have been reported as well. It is fascinating to hear the story of Daryl Davis, an African American blues musician, who has personally converted over 300 former Ku Klux Klansmen, simply by befriending and talking to them. Mark Segraves, "'How Can You Hate Me?' Maryland Musician Converts White Supremacists," *NBC Washington*, February 14, 2020, https://www .nbcwashington.com/news/local/musician-fights-racism-by-speaking-to-white -supremacists/2216483/. For a video interview, with Davis see http://www.pbs.org /wnet/amanpour-and-company/video/daryl-davis-on-befriending-members-of-the -kkk/.

51. Eli Saslow, *Rising Out of Hatred: The Awakening of a Former White Nationalist* (New York: Anchor, 2018), 225.

52. Charles Monroe-Kane, "Can You Change the Mind of a White Supremacist?" *To the Best of Our Knowledge*, March 12, 2019, https://www.ttbook.org/interview/can -you-change-mind-white-supremacist.

53. David Weissman, "I Used to Be a Trump Troll—Until Sarah Silverman Engaged with Me," *Forward*, June 5, 2018, https://forward.com/scribe/402478/i-was-a-trump -troll/; "Former Twitter Troll Credits Sarah Silverman with Helping Him to See 'How Important Talking Is,'" *CBC Radio*, April 12, 2019, https://www.cbc.ca/radio /outintheopen/switching-sides-1.5084481/former-twitter-troll-credits-sarah-silverman -with-helping-him-see-how-important-talking-is-1.5094232; Jessica Kwong, "Former Trump Supporter Says MAGA 'Insults' Snapped Him Out of 'Trance' of Supporting President," *Newsweek*, October 3, 2019, https://www.newsweek.com/former-trump -supporter-snapped-out-maga-1463021.

54. Michael B. Kelley, "STUDY: Watching Only Fox News Makes You Less Informed Than Watching No News At All," *Business Insider*, May 22, 2012, https://www .businessinsider.com/study-watching-fox-news-makes-you-less-informed-than -watching-no-news-at-all-2012-5.

55. Vanessa Milne, "Seven Ways to Talk to Anti-vaxxers (That Might Actually Change Their Minds)," *Healthy Debate*, August 31, 2017, https://healthydebate.ca /2017/08/topic/vaccine-safety-hesitancy.

56. Karin Kirk, "How to Identity People Open to Evidence about Climate Change," *Yale Climate Connection*, November 9, 2018, https://www.yaleclimateconnections .org/2018/11/focus-on-those-with-an-open-mind/.

57. John M. Glionna, "The Real-Life Conversion of a Former Anti-Vaxxer," *California Healthline*, August 2, 2019, https://californiahealthline.org/news/the-real-life -conversion-of-a-former-anti-vaxxer/.

58. Lewandowsky and Cook, *The Conspiracy Theory Handbook*.

59. Rene Chun, "Scientists Are Trying to Figure Out Why People Are OK With Trump's Endless Supply of Lies," *Los Angeles Magazine*, November 14, 2019, https://www.lamag.com/citythinkblog/trump-lies-research/.

60. Christopher Joyce, "Rising Seas Made This Republican Mayor a Climate Change Believer," *NPR*, May 17, 2016, https://www.npr.org/2016/05/17/477014145/rising-seas-made-this-republican-mayor-a-climate-change-believer.

61. Fred Grimm, "Florida's Mayors Face Reality of Rising Seas and Climate Change," *Miami Herald*, March 14, 2016, https://www.miamiherald.com/news/local/news-columns-blogs/fred-grimm/article68092452.html.

62. Sarah Ann Wheeler and Celine Nauges, "Farmers' Climate Denial Begins to Wane as Reality Bites," *The Conversation*, October 11, 2018, https://theconversation.com/farmers-climate-denial-begins-to-wane-as-reality-bites-103906; Helena Bottemiller Evich, "'I'm Standing Right Here in the Middle of Climate Change': How USDA Is Failing Farmers," *Politico*, October 15, 2019, https://www.politico.com/news/2019/10/15/im-standing-here-in-the-middle-of-climate-change-how-usda-fails-farmers-043615; "Stories from the Sea: Fishermen Confront Climate Change," *Washington Nature*, https://www.washingtonnature.org/fishermen-climate-change.

63. Emma Reynolds, "Some Anti-vaxxers Are Changing Their Minds Because of the Coronavirus Pandemic," *CNN*, April 20, 2020, https://www.cnn.com/2020/04/20/health/anti-vaxxers-coronavirus-intl/index.html; Jon Henley, "Coronavirus Causing Some Anti-vaxxers to Waver, Experts Say," *Guardian*, April 21, 2020, https://www.theguardian.com/world/2020/apr/21/anti-vaccination-community-divided-how-respond-to-coronavirus-pandemic; Victoria Waldersee, "Could the New Coronavirus Weaken 'Anti-vaxxers'?" *Reuters*, April 11, 2020, https://www.reuters.com/article/us-health-coronavirus-antivax/could-the-new-coronavirus-weaken-anti-vaxxers-idUSKCN21T089.

Chapter 4

1. Chris Mooney and Brady Dennis, "The World Has Just Over a Decade to Get Climate Change under Control," *Washington Post*, October 7, 2018, https://www.washingtonpost.com/energy-environment/2018/10/08/world-has-only-years-get-climate-change-under-control-un-scientists-say/; Nina Chestney, "Global Carbon Emissions Hit Record High in 2017," *Reuters*, March 22, 2018, https://www.reuters.com/article/us-energy-carbon-iea/global-carbon-emissions-hit-record-high-in-2017-idUSKBN1GY0RB.

2. Brady Dennis and Chris Mooney, "'We Are in Trouble': Global Carbon Emissions Reached a Record High in 2018," *Washington Post*, December 5, 2018, https://www

.washingtonpost.com/energy-environment/2018/12/05/we-are-trouble-global
-carbon-emissions-reached-new-record-high/; Damian Carrington, "'Brutal News':
Global Carbon Emissions Jump to All Time High in 2018," *Guardian,* December 5,
2018, https://www.theguardian.com/environment/2018/dec/05/brutal-news-global
-carbon-emissions-jump-to-all-time-high-in-2018; Brad Plumer, "U.S. Carbons Emissions Surged in 2018, Even as Coal Plants Closed," *New York Times,* January 8, 2019,
https://www.nytimes.com/2019/01/08/climate/greenhouse-gas-emissions-increase
.html.

3. Chelsea Harvey and Nathanial Gronewold, "CO$_2$ Emissions Will Break Another
Record in 2019," *Scientific American,* December 4, 2019, https://www.scientificamerican
.com/article/co2-emissions-will-break-another-record-in-2019/. Two pieces of good
news, though: although global 2019 emissions are expected to be the highest ever,
the rate of growth appears to be declining. Chelsea Harvey and Nathanial Gronewold,
"Greenhouse Gas Emissions to Set New Record This Year, but Rate of Growth Shrinks,"
Science, December 4, 2019, https://www.sciencemag.org/news/2019/12/greenhouse
-gas-emissions-year-set-new-record-rate-growth-shrinks. In the US, greenhouse gas
emissions actually fell 2.1 percent in 2019, due largely to decreased coal consumption. Unfortunately, this still puts us far off the pace to reach our pledge through
the Paris Agreement. Steve Mufson, "U.S. Greenhouse Gas Emissions Fell Slightly
in 2019," *Washington Post,* January 7, 2020, https://www.washingtonpost.com/climate
-environment/us-greenhouse-gas-emissions-fell-slightly-in-2019/2020/01/06/568f0a82
-309e-11ea-a053-dc6d944ba776_story.html.

4. Dennis and Mooney, "We Are in Trouble."

5. Coral Davenport, "Major Climate Report Describes a Strong Risk of Crisis as Early
as 2040," *New York Times,* October 7, 2018, https://www.nytimes.com/2018/10/07
/climate/ipcc-climate-report-2040.html; Dennis and Mooney, "We Are in Trouble";
Emily Holden, "'It'll Change Back': Trump Says Climate Change Not a Hoax, but
Denies Lasting Impact," *Guardian,* October 15, 2018, https://www.theguardian.com
/us-news/2018/oct/15/itll-change-back-trump-says-climate-change-not-a-hoax-but
-denies-lasting-impact.

6. Brady Dennis, "In Bleak Report, UN Says Drastic Action Is Only Way to Avoid
Worst Impacts of Climate Change," *Washington Post,* November 26, 2019, https://
www.washingtonpost.com/climate-environment/2019/11/26/bleak-report-un
-says-drastic-action-is-only-way-avoid-worst-impacts-climate-change/; Alister Doyle,
"Global Warming May Be More Severe Than Expected by 2100: Study," *Reuters,*
December 6, 2017, https://www.reuters.com/article/us-climatechange-temperatures
/global-warming-may-be-more-severe-than-expected-by-2100-study-idUSKBN1E02J6;
Dave Mosher and Aylin Woodward, "What Earth Might Look Like in 80 Years if
We're Lucky—and if We're Not," *Business Insider,* October 17, 2019, https://www
.businessinsider.com/paris-climate-change-limits-100-years-2017-6; Jen Christensen

and Michael Nedelman, "Climate Change Will Shrink U.S. Economy and Kill Thousands, Government Report Warns," *CNN*, November 26, 2018, https://www.cnn.com/2018/11/23/health/climate-change-report-bn/index.html.

7. Coral Davenport, "Major Climate Report Describes a Strong Risk of Crisis," *New York Times*, October 7, 2018, https://www.nytimes.com/2018/10/07/climate/ipcc-climate-report-2040.html.

8. Davenport, "Major Climate Report"; Ron Meador, "New Outlook on Global Warming: Best Prepare for Social Collapse, and Soon," *Minnpost*, October 15, 2018, https://www.minnpost.com/earth-journal/2018/10/new-outlook-on-global-warming-best-prepare-for-social-collapse-and-soon/.

9. Paul Bledsoe, "Going Nowhere Fast and Climate Change, Year After Year," *New York Times*, December 19, 2018, https://www.nytimes.com/2018/12/29/opinion/climate-change-global-warming-history.html; Dennis and Mooney, "We Are in Trouble."

10. Mooney and Dennis, "The World Has Just Over a Decade to Get Climate Change under Control."

11. Dennis and Mooney, "We Are in Trouble"; Mooney and Dennis, "The World Has Just Over a Decade to Get Climate Change under Control."

12. Steven Mufson, "'A Kind of Dark Realism': Why the Climate Change Problem Is Starting to Look Too Big to Solve," *Washington Post*, December 4, 2018, https://www.washingtonpost.com/national/health-science/a-kind-of-dark-realism-why-the-climate-change-problem-is-starting-to-look-too-big-to-solve/2018/12/03/378e49e4-e75d-11e8-a939-9469f1166f9d_story.html.

13. Doyle Rice, "Coal Is the Main Offender for Global Warming, and Yet the World Is Using It More Than Ever," *USA Today*, March 26, 2019, https://www.usatoday.com/story/news/nation/2019/03/26/climate-change-coal-still-king-global-carbon-emissions-soar/3276401002/.

14. Dennis and Mooney, "We Are in Trouble."

15. Simon Carraud and Michel Rose, "Macron Makes U-turn on Fuel Tax Increase, in Face of 'Yellow Vest' Protests," *Reuters*, December 4, 2018, https://www.reuters.com/article/us-france-protests/macron-makes-u-turn-on-fuel-tax-increases-in-face-of-yellow-vest-protests-idUSKBN1O30MX.

16. Mufson, "A Kind of Dark Realism."

17. Brady Dennis, "Trump Makes It Official: U.S. Will Withdraw from Paris Climate Accords," *Washington Post*, November 4, 2019, https://www.washingtonpost.com/climate-environment/2019/11/04/trump-makes-it-official-us-will-withdraw-paris-climate-accord/.

18. David Roberts, "The Trump Administration Just Snuck through Its Most Devious Coal Subsidy Yet," *Vox*, December 23, 2019, https://www.vox.com/energy-and-environment/2019/12/23/21031112/trump-coal-ferc-energy-subsidy-mopr.

19. Nathan Rott and Jennifer Ludden, "Trump Administration Weakens Auto Emissions Standards," *NPR*, March 31, 2020, https://www.npr.org/2020/03/31/824431240/trump-administration-weakens-auto-emissions-rolling-back-key-climate-policy.

20. Patrick Kingsley, "Trump Says California Can Learn from Finland on Fires. Is He Right?" *New York Times*, November 18, 2018, https://www.nytimes.com/2018/11/18/world/europe/finland-california-wildfires-trump-raking.html.

21. Jennifer Rubin, "Trump Shows the Rank Dishonesty of Climate-Change Deniers," *Washington Post*, October 15, 2018, https://www.washingtonpost.com/news/opinions/wp/2018/10/15/trump-shows-the-rank-dishonesty-of-climate-change-deniers/.

22. Josh Dawsey et al, "Trump on Climate Change: 'People Like Myself, We Have Very High Levels of Intelligence but We're Not Necessarily Such Believers,'" *Washington Post*, November 27, 2018, https://www.washingtonpost.com/politics/trump-on-climate-change-people-like-myself-we-have-very-high-levels-of-intelligence-but-were-not-necessarily-such-believers/2018/11/27/722f0184-f27e-11e8-aeea-b85fd44449f5_story.html; Matt Viser, "'Just a Lot of Alarmism': Trump's Skepticism of Climate Science Is Echoed across GOP," *Washington Post*, December 2, 2018, https://www.washingtonpost.com/politics/just-a-lot-of-alarmism-trumps-skepticism-of-climate-science-is-echoed-across-gop/2018/12/02/f6ee9ca6-f4de-11e8-bc79-68604ed88993_story.html.

23. Brady Dennis and Chris Mooney, "Major Trump Administration Climate Report Says Damage Is 'Intensifying across the Country,'" *Washington Post*, November 23, 2018, https://www.washingtonpost.com/energy-environment/2018/11/23/major-trump-administration-climate-report-says-damages-are-intensifying-across-country/; Jen Christensen and Michael Nedelman, "Climate Change Will Shrink U.S. Economy and Kill Thousands," *CNN*, November 23, 2018, https://www.cnn.com/2018/11/23/health/climate-change-report-bn/index.html.

24. There is some controversy over bold claims that global warming will render the planet uninhabitable or lead to the extinction of our species. Nonetheless, there is no doubt that it would drastically alter human life on this planet, leading to untold misery and millions of deaths. This is not to mention the issue of mass extinction and loss of biodiversity of other species. Robert Watson, "Loss of Biodiversity Is Just as Catastrophic as Climate Change," *Guardian*, May 6, 2019, https://www.theguardian.com/commentisfree/2019/may/06/biodiversity-climate-change-mass-extinctions; Michael Shellenberger, "Why Apocalyptic Claims About Climate Change Are Wrong," *Forbes*, November 25, 2019, https://www.forbes.com/sites/michaelshellenberger/2019/11/25/why-everything-they-say-about-climate-change-is-wrong/#5cea81cb12d6; Chris Mooney, "Scientists Challenge Magazine

Story about 'Uninhabitable Earth," *Washington Post*, July 12, 2017, https://www
.washingtonpost.com/news/energy-environment/wp/2017/07/12/scientists
-challenge-magazine-story-about-uninhabitable-earth/; Jen Christensen, "250,000
Deaths a Year from Climate Change Is a 'Conservative Estimate,' Research Says,"
CNN, January 16, 2019, https://www.cnn.com/2019/01/16/health/climate-change
-health-emergency-study/index.html; "The Impact of Global Warming on Human
Fatality Rates," *Scientific American*, June 7, 2009, https://www.scientificamerican
.com/article/global-warming-and-health/.

25. "Climate Concerns Increase: Most Republicans Now Acknowledge Change,"
Monmouth, November 29, 2018, https://www.monmouth.edu/polling-institute/reports
/monmouthpoll_us_112918/. A number of other recent polls confirm these results as
well: John Schwartz, "Global Warming Concerns Rise Among Americans in New Poll,"
New York Times, January 22, 2019, https://www.nytimes.com/2019/01/22/climate
/americans-global-warming-poll.html; Robinson Meyer, "Voters Really Care about
Climate Change," *Atlantic*, February 21, 2020, https://www.theatlantic.com/science
/archive/2020/02/poll-us-voters-really-do-care-about-climate-change/606907/; Brady
Dennis et al., "Americans Increasingly See Climate Change as a Crisis, Poll Shows,"
Washington Post, September 13, 2019, https://www.washingtonpost.com/climate
-environment/americans-increasingly-see-climate-change-as-a-crisis-poll-shows/2019
/09/12/74234db0-cd2a-11e9-87fa-8501a456c003_story.html; "Scientific Consensus:
Earth's Climate Is Warming," NASA, https://climate.nasa.gov/scientific-consensus/.

26. See chapter 7 of Chris Mooney, *The Republican War on Science* (New York: Basic
Books, 2005).

27. Two of the best are James Hansen, *Storms of My Grandchildren* (New York:
Bloomsbury, 2009) and James Hoggan, *Climate Cover-Up: The Crusade to Deny Global
Warming* (Vancouver: Greystone, 2009).

28. Mooney, *The Republican War on Science*, 81.

29. James Lawrence Powell, "Why Climate Deniers Have No Scientific Credibility—in
One Pie Chart," *Desmog*, November 15, 2012, https://www.desmogblog.com/2012
/11/15/why-climate-deniers-have-no-credibility-science-one-pie-chart.

30. James Lawrence Powell: "Why Climate Deniers Have No Scientific Credibility:
Only 1 of 9,136 Recent Peer-Reviewed Authors Rejects Global Warming," *Desmog*,
January 8, 2014, https://www.desmogblog.com/2014/01/08/why-climate-deniers
-have-no-scientific-credibility-only-1-9136-study-authors-rejects-global-warming.

31. Peter T. Doran and Maggie Kendall Zimmerman, "Examining the Scientific Consensus on Climate Change," *Eos: Transactions of the American Geophysical Union* 90,
no. 3 (2009): 22–23.

32. John Cook et al., "Quantifying the Consensus on Anthropogenic Global
Warming in the Scientific Literature," *Environmental Research Letters*, May 15, 2013,

https://iopscience.iop.org/article/10.1088/1748-9326/8/2/024024/pdf; "The 97% Consensus on Global Warming," *Skeptical Science*, https://www.skepticalscience.com/global-warming-scientific-consensus-intermediate.htm.

33. Katherine Ellen Foley, "Those 3% of Scientific Papers That Deny Climate Change? A Review Found Them All Flawed," *Quartz*, September 5, 2017, https://qz.com/1069298/the-3-of-scientific-papers-that-deny-climate-change-are-all-flawed/; Dana Nuccitelli, "Millions of Times Later, 97 Percent Climate Consensus Still Faces Denial," *Bulletin of the Atomic Scientists*, August 15, 2019, https://thebulletin.org/2019/08/millions-of-times-later-97-percent-climate-consensus-still-faces-denial/; Dana Nuccitelli, "Here's What Happens When You Try to Replicate Climate Contrarian Papers," *Guardian*, August 25, 2015, https://www.theguardian.com/environment/climate-consensus-97-per-cent/2015/aug/25/heres-what-happens-when-you-try-to-replicate-climate-contrarian-papers; Rasmus E. Benestad et al., "Learning from Mistakes in Climate Research," *Theoretical and Applied Climatology* 126 (2016): 699–703, https://link.springer.com/article/10.1007/s00704-015-1597-5.

34. It is stunning to note that even today, 150 years after Darwin's breakthrough on evolution by natural selection (which is the backbone for *all* biological explanation), there is still only 98 percent consensus among scientists. David Masci, "For Darwin Day, 6 Facts about the Evolution Debate," Pew Research Center's Fact Tank, February11, 2019, https://www.pewresearch.org/fact-tank/2019/02/11/darwin-day/.

35. Dana Nuccitelli, "Trump Thinks Scientists Are Split on Climate Change. So Do Most Americans," *Guardian*, October 22, 2018, https://www.theguardian.com/environment/climate-consensus-97-per-cent/2018/oct/22/trump-thinks-scientists-are-split-on-climate-change-so-do-most-americans.

36. It was created just after the Kyoto Protocol in December 1997 and leaked in April 1998. "1998 American Petroleum Institute Global Climate Science Communications Team Action Plan," *Climate Files*, http://www.climatefiles.com/trade-group/american-petroleum-institute/1998-global-climate-science-communications-team-action-plan/.

37. "Climate Science vs. Fossil Fuel Fiction," Union of Concerned Scientists, March 2015, https://www.ucsusa.org/sites/default/files/attach/2015/03/APIquote1998_1.pdf.

38. Shannon Hall, "Exxon Knew about Climate Change almost 40 Years Ago," *Scientific American*, October 26, 2015, https://www.scientificamerican.com/article/exxon-knew-about-climate-change-almost-40-years-ago/; Suzanne Goldenberg, "Exxon Knew of Climate Change in 1981, Email Says—But It Funded Deniers for 27 More Years," *Guardian*, July 8, 2015, https://www.theguardian.com/environment/2015/jul/08/exxon-climate-change-1981-climate-denier-funding.

39. It has always mystified me that adherents of conspiracy theories are not more interested in all this—a genuine conspiracy! For more, see Steve Coll, *Private Empire:*

ExxonMobil and American Power (New York: Penguin Press, 2012); "ExxonMobil: A 'Private Empire' on the World Stage," *NPR*, May 2, 2012, http://www.npr.org/2012 /05/02/151842205/exxonmobil-a-private-empire-on-the-world-stage.

40. Naomi Oreskes and Erik Conway, *Merchants of Doubt* (New York: Bloomsbury, 2011), 183.

41. "Nancy Pelosi and Newt Gingrich Commercial on Climate Change," YouTube, uploaded April 17, 2008, https://www.youtube.com/watch?v=qi6n_-wB154.

42. In *Merchants of Doubt*, Oreskes and Conway argue that "an organized campaign of denial began [in 1989], and soon ensnared the entire climate science community" (183). This eventually led public opinion on the reality of global warming to *decline* over time, even as the scientists grew more certain, especially after a pressure campaign on journalists to give equal time to the "other side" of the climate change "debate" (169–170, 214–215). By the time this reached Congress, any real hope of fighting climate change was dead. In 1997, the Senate voted to block America's adoption of the Kyoto Protocol (215).

43. Jane Mayer, *Dark Money: The Hidden History of the Billionaires behind the Rise of the Radical Right* (New York: Doubleday, 2016).

44. Mayer, *Dark Money*, 204.

45. Jane Mayer, "'Kochland' Examines the Koch Brothers' Early, Crucial Role in Climate-Change Denial," *New Yorker*, August 13, 2019, https://www.newyorker .com/news/daily-comment/kochland-examines-how-the-koch-brothers-made-their -fortune-and-the-influence-it-bought. See also Christopher Leonard, *Kochland: The Secret History of Koch Industries and Corporate Power in America* (New York: Simon and Schuster, 2019).

46. Emily Atkin, "How David Koch Change the World," *New Republic*, August 23, 2019, https://newrepublic.com/article/154836/david-koch-changed-world.

47. Niall McCarthy, "Oil and Gas Giants Spend Millions Lobbying to Block Climate Change Policies," *Forbes*, March 25, 2019, https://www.forbes.com/sites /niallmccarthy/2019/03/25/oil-and-gas-giants-spend-millions-lobbying-to-block -climate-change-policies-infographic/#5c28b08c7c4f.

48. Colin Schultz, "Meet the Money Behind the Climate Denial Movement," *Smithsonian*, December 23, 2013, https://www.smithsonianmag.com/smart-news/meet -the-money-behind-the-climate-denial-movement-180948204/.

49. Justin Gillis and Leslie Kaufman, "Leaks Offer Glimpse of Campaign Against Climate Change," *New York Times*, February 15, 2012, https://www.nytimes.com /2012/02/16/science/earth/in-heartland-institute-leak-a-plan-to-discredit-climate -teaching.html; Gayathri Vaidyanathan, "Think Tank That Cast Doubt on Climate

Change Science Morphs into Smaller One," *E&E News*, December 10, 2015, https://www.eenews.net/stories/1060029290.

50. Brendan Montague, "I Crashed a Climate Change Denial Conference in Las Vegas," *Vice*, July 22, 2014, https://www.vice.com/da/article/7bap4x/las-vegas-climate-change-denial-brendan-montague-101; Brian Palmer, "What It's Like to Attend a Climate Denial Conference," *Pacific Standard*, December 16, 2015, https://psmag.com/environment/what-its-like-to-attend-a-climate-denial-conference.

51. Mayer, *Dark Money*, 213.

52. *Dark Money*, 214. See also Oreskes and Conway, *Merchants of Doubt*, 169–170. After creeping up for several years, public polling on climate change suddenly started to go down.

53. *Dark Money*, 211.

54. Atkin, "How David Koch Changed the World."

55. Marc Morano, spokesman for Senator James Inhofe (R-Oklahoma): "Gridlock is the greatest friend a global warming skeptic has, because that's all you really want. . . . There's no legislation we're championing. We're the negative force. We are just trying to stop stuff" (Mayer, *Dark Money*, 224–225)

56. David Roberts, "Why Conservatives Keep Gaslighting the Nation about Climate Change," *Vox*, October 31, 2018, https://www.vox.com/energy-and-environment/2018/10/22/18007922/climate-change-republicans-denial-marco-rubio-trump.

57. Nadja Popovich, "Climate Change Rises as a Public Priority, but It's More Partisan Than Ever," *New York Tiimes*, February 20, 2020, https://www.nytimes.com/interactive/2020/02/20/climate/climate-change-polls.html; Brian Kennedy, "U.S. Concern about Climate Change Is Rising, but Mainly among Democrats," Pew Research, April 16, 2020, https://www.pewresearch.org/fact-tank/2020/04/16/u-s-concern-about-climate-change-is-rising-but-mainly-among-democrats/.

58. In a different Pew poll, done on Earth Day 2020, it was found that "partisanship is a stronger factor in people's beliefs about climate change than is their level of knowledge and understanding of science." Gary Funk and Brian Kennedy, "How Americans Sees Climate Change and the Environment in 7 Charts," Pew Research, April 21, 2020, https://www.pewresearch.org/fact-tank/2020/04/21/how-americans-see-climate-change-and-the-environment-in-7-charts/.

59. Alister Doyle, "Evidence for Man-Made Global Warming Hits Gold Standard," *Reuters*, February 25, 2019, https://www.reuters.com/article/us-climatechange-temperatures/evidence-for-man-made-global-warming-hits-gold-standard-scientists-idUSKCN1QE1ZU.

60. Atkin, "How David Koch Changed the World."

61. Chris Mooney, "Ted Cruz Keeps Saying That Satellites Don't Show Global Warming. Here's the Problem," *Washington Post*, January 29, 2016, https://www .washingtonpost.com/news/energy-environment/wp/2016/01/29/ted-cruz-keeps -saying-that-satellites-dont-show-warming-heres-the-problem/; Lauren Carroll, "Ted Cruz's World's on Fire, but Not for the Last 17 Years," *Politifact*, March 20, 2015, https://www.politifact.com/factchecks/2015/mar/20/ted-cruz/ted-cruzs-worlds-fire -not-last-17-years/.

62. Jeremy Schulman, "Every Insane Thing Donald Trump Has Said about Global Warming," *Mother Jones*, December 12, 2018, https://www.motherjones.com/environ ment/2016/12/trump-climate-timeline/.

63. Kate Sheppard, "Climategate: What Really Happened?" *Mother Jones*, April 21, 2011, https://www.motherjones.com/environment/2011/04/history-of-climategate/.

64. "What If You Held a Conference, and No (Real) Scientists Came?" *RealClimate*, January 30, 2008, http://www.realclimate.org/index.php/archives/2008/01/what -if-you-held-a-conference-and-no-real-scientists-came/comment-page-8/; Brendan Montague, "I Crashed a Climate Change Denial Conference in Last Vegas," *Vice*, July 22, 2014, https://www.vice.com/en/article/7bap4x/las-vegas-climate-change-denial -brendan-montague-101; https://climateconference.heartland.org/. See my discussion of climate change in *Respecting Truth* (72–80).

65. "How Do We Know That Humans Are the Major Cause of Global Warming?" Union of Concerned Scientists, July 14, 2009 (updated August 1, 2017), https://www .ucsusa.org/resources/are-humans-major-cause-global-warming; "CO_2 Is Main Driver of Climate Change," *Skeptical Science*, July 15, 2015, https://www.skepticalscience .com/CO2-is-not-the-only-driver-of-climate.htm.

66. Chris Mooney and Elise Viebeck, "Trump's Economic Advisor and Marco Rubio Question Extent of Human Contribution to Climate Change," *Washington Post*, October 14, 2018, https://www.washingtonpost.com/powerpost/larry-kudlow -marco-rubio-question-extent-of-human-contribution-to-climate-change/2018/10 /14/c8606ae2-cfcf-11e8-b2d2-f397227b43f0_story.html; Alan Yuhas, "Republicans Reject Climate Change Fears Despite Rebukes from Scientists," *Guardian*, February 1, 2016, https://www.theguardian.com/us-news/2016/feb/01/republicans-ted-cruz -marco-rubio-climate-change-scientists.

67. Alister Doyle and Bruce Wallace, "U.N. Climate Deal in Paris May Be Graveyard for 2C Goal," Reuters, June 1, 2015, https://www.reuters.com/article/us-climatechange -paris-insight/u-n-climate-deal-in-paris-may-be-graveyard-for-2c-goal-idUSKBN0OH1G 820150601.

68. Jon Henley, "The Last Days of Paradise," *Guardian*, November 10, 2008, https:// www.theguardian.com/environment/2008/nov/11/climatechange-endangered -habitats-maldives.

69. Nudity or religious proselytizing can lead to arrest in the Maldives. Pornography and idols of worship are both banned.

70. This actually happened in 2004, when a one-meter-high wave hit one-meter-high Malé and killed eighty-two people, displacing 12,000 more.

71. There is actually a third danger, already mentioned, though this one is to human ability to thrive on the island rather than to the island itself. As overwash becomes more common, the island's fresh-water supply gets contaminated. When the storms happen more often, and occur earlier in the season, pretty soon there is not enough fresh water to sustain the human population. Josh Gabbatiss, "Rising Sea Levels Could Make Thousands of Islands from the Maldives to Hawaii 'Uninhabitable within Decades,'" *Independent*, April 25, 2018, https://www.independent.co.uk/environment/islands-sea-level-rise-flooding-uninhabitable-climate-change-maldives-seychelles-hawaii-a8321876.html.

72. It's a little known fact, but the US has its own version of the Maldives, called Tangier Island, out in Chesapeake Bay. As the water rises, its marshland is sinking and will eventually turn into open water. Ironically, the residents of Tangier Island consist of an inordinate number of climate deniers and Trump supporters. Eventually, they will be the first climate refugees in the United States. David J. Unger, "On a Sinking Island, Climate Science Takes a Back Seat to the Bible," *Grist*, September 3, 2018, https://grist.org/article/on-a-sinking-island-climate-science-takes-a-back-seat-to-the-bible/; Simon Worrall, "Tiny U.S. Island Is Drowning. Residents Deny the Reason," *National Geographic*, September 7, 2018, https://www.nationalgeographic.com/environment/2018/09/climate-change-rising-seas-tangier-island-chesapeake-book-talk/.

73. This is precisely the kind of "ground truth" experience that the Flat Earth deniers had demanded. "Have you ever been on a spaceship?" "Have you ever flown over Antarctica?" "Have you ever been out on a boat sixty miles from Chicago?" Now I can tell them, "No, but I did go to the other side of the world, where I saw the effects of climate change . . . and some different stars."

74. Just after we got back from the Maldives, I read about some hope for artificial mechanisms to maintain the islands, even as the coral dies. MIT scientists are working to try to place underwater bladders to catch the sand and build up artificial reefs to keep the islands from breaking apart. Courtney Linder, "The Extraordinary Way We'll Rebuild Our Shrinking Islands," *Popular Mechanics*, May 25, 2020, https://www.popularmechanics.com/science/green-tech/a32643071/rebuilding-islands-ocean-waves/.

Chapter 5

1. Calvin Woodward and Seth Borenstein, "Unraveling the Mystery of whether Cows Fart," *AP*, April 28, 2019, https://apnews.com/9791f1f85808409e93a1abc8b98531d5.

2. "Sources of Greenhouse Gas Emissions," EPA, https://www.epa.gov/ghgemissions /sources-greenhouse-gas-emissions.

3. Tess Riley, "Just 100 Companies Responsible for 71% of Global Emissions, Study Says," *Guardian*, July 10, 2017, https://www.theguardian.com/sustainable-business /2017/jul/10/100-fossil-fuel-companies-investors-responsible-71-global-emissions -cdp-study-climate-change.

4. See the section "The Origins and Causes of Climate Denial" in chapter 4.

5. Robinson Meyer, "America's Coal Consumption Entered Free Fall in 2019," *Atlantic*, January 7, 2020, https://www.theatlantic.com/science/archive/2020/01/americas -coal-consumption-entered-free-fall-2019/604543/#:~:text=American%20coal%20 use%20fell%2018,is%20remarkable%2C%E2%80%9D%20Houser%20said.

6. Unfortunately, this does not mean that US emissions have dropped proportionately. The US continues to be the second largest emitter of carbon dioxide, as we have replaced coal with natural gas, another fossil fuel. We are still projected to miss our target of 26 percent emissions reduction by 2025, as outlined in the Paris Agreement.

7. Somini Sengupta, "The World Needs to Quit Coal. Why Is It So Hard?" *New York Times*, November 24, 2018, https://www.nytimes.com/2018/11/24/climate/coal-global -warming.html.

8. "The Road to a Paris Climate Deal," *New York Times*, December 11, 2015, https:// www.nytimes.com/interactive/projects/cp/climate/2015-paris-climate-talks/where -in-the-world-is-climate-denial-most-prevalent.

9. In a more recent survey, it was found that out of twenty-three large countries, only Saudi Arabia and Indonesia had a higher proportion of climate deniers than the US. Oliver Milman and Fiona Harvey, "US Is Hotbed of Climate Change Denial, Major Global Survey Finds," *Guardian*, May 8, 2019, https://www.theguardian.com /environment/2019/may/07/us-hotbed-climate-change-denial-international-poll.

10. Chris Mooney, "The Strange Relationship between Global Warming Denial and . . . Speaking English," *Guardian*, July 23, 2014, https://www.theguardian.com /environment/2014/jul/23/the-strange-relationship-between-global-warming-denial -and-speaking-english After the US, the next worst were the UK and Australia.

11. Greed? Self-concern? Indifference? Maybe that kid on the boat in the Maldives had it right. It's not just about belief, it's about caring.

12. Brad Plumer, "Carbon Dioxide Emissions Hit a Record in 2019, Even as Coal Fades," *New York Times*, December 3, 2019, https://www.nytimes.com/2019/12/03 /climate/carbon-dioxide-emissions.html.

13. "Energy and the Environment Explained," US Energy Information Administration, last updated August 11, 2020, https://www.eia.gov/energyexplained/energy-and -the-environment/where-greenhouse-gases-come-from.php.

14. "Salem-Style Mass Hysteria Animates Trump Movement at Moon Twp., PA Rally—Nov. 6, 2016," https://www.youtube.com/watch?v=BQNmjpXBanc&t=4s; "Creating Breakthrough Moments," https://www.youtube.com/watch?v=OOfV4ZkmjlM; "Trump Voter Breakthrough—May 2017," https://www.youtube.com/watch?v=7V8JZXx_hUs.

15. Unfortunately, due to a scheduling complication, the NPR show couldn't air until after the meeting had taken place, but I still enjoyed talking to the host about climate denial. Kara Holsopple, "The Philosophy of Climate Denial," *Allegheny Front*, September 18, 2019, https://www.alleghenyfront.org/the-philosophy-of-climate-denial/.

16. https://quoteinvestigator.com/2017/11/30/salary/.

17. Eliza Griswold, "People in Coal Country Worry about the Climate, Too," *New York Times*, July 13, 2019, https://www.nytimes.com/2019/07/13/opinion/sunday/jobs-climate-green-new-deal.html.

18. Remember that 18 percent drop in coal in 2019? This is what he was talking about.

19. Jane Mayer, *Dark Money: The Hidden History of the Billionaires Behind the Rise of the Radical Right* (New York: Anchor, 2017).

20. Andrew Norman, *Mental Immunity: Infectious Ideas, Mind-Parasites, and the Search for a Better Way to Think* (New York: Harper Wave, 2021).

21. All of these names are pseudonyms.

22. Griswold, "People in Coal Country Worry about the Climate, Too."

23. Jake Johnson, "'We Are in a Climate Emergency, America': Anchorage Hits 90 Degrees for the First Time in Recorded History," *Common Dreams*, July 5, 2019, https://www.commondreams.org/news/2019/07/05/we-are-climate-emergency-america-anchorage-hits-90-degrees-first-time-recorded.

24. Alejandra Borunda, "What a 100-Degree Day in Siberia Really Means," *National Geographic*, June 23, 2020, https://www.nationalgeographic.com/science/2020/06/what-100-degree-day-siberia-means-climate-change/

25. I, of course, am not the first person to notice the obvious similarities here. Indeed, since all science denial is basically the same, it is notable that the type of denial we find for coronavirus is eerily similar to that for climate change and other issues. Misinformation, wishful thinking, blame, denial, and making up facts, followed by the claim that this is just too expensive to solve, is right out of the science denier's playbook. Katelyn Weisbrod, "6 Ways Trump's Denial of Science Has Delayed the Response to COVID-19 (and Climate Change)," *Inside Climate News*, March 19, 2020, https://insideclimatenews.org/news/19032020/denial-climate-change-coronavirus-donald-trump?gclid=EAIaIQobChMIsan_qduf6gIVDo3ICh1XPAIuEAAYASAAEgID 0_D_BwE; Gilad Edelman, "The Analogy between COVID-19 and Climate Change Is Eerily Precise," *Wired*, March 25, 2020, https://www.wired.com/story/the-analogy-between-covid-19-and-climate-change-is-eerily-precise/.

26. Unfortunately, there is no vaccine for climate change.

27. Chris Mooney, Brady Dennis, and John Muyskens, "Global Emissions Plunged an Unprecedented 17% during the Coronavirus Pandemic," *Washington Post*, May 19, 2020, https://www.washingtonpost.com/climate-environment/2020/05/19/greenhouse -emissions-coronavirus.

28. Brad Plumber, "Emissions Declines Will Set Records This Year. But It's Not Good News," *New York Times*, April 30, 2020, https://www.nytimes.com/2020/04/30 /climate/global-emissions-decline.html.

29. Maggie Haberman and David Sanger, "Trump Says Coronavirus Cure 'Cannot Be Worse Than the Problem Itself,'" *New York Tiimes*, March 23, 2020, https://www .nytimes.com/2020/03/23/us/politics/trump-coronavirus-restrictions.html.

30. Lois Beckett, "Older People Would Rather Die Than Let Covid-19 Harm the US Economy—Texas Official," *Guardian*, March 24, 2020, https://www.theguardian.com /world/2020/mar/24/older-people-would-rather-die-than-let-covid-19-lockdown -harm-us-economy-texas-official-dan-patrick; Sally Jenkins, "Some May Have to Die to Save the Economy? How about Offering Testing and Basic Protections?" *Washington Post*, April 18, 2020, https://www.washingtonpost.com/sports/2020/04/18/sally -jenkins-trump-coronavirus-testing-economy/.

31. Indeed, a poll taken just after the 2016 presidential election showed that 62 percent of Trump voters supported carbon pollution taxes! If the voters are there, why aren't the politicians? Dana Nuccitelli, "Trump Can Save His Presidency with a Great Deal to Save the Climate," *Guardian*, February 22, 2017, https://www.theguardian .com/environment/climate-consensus-97-per-cent/2017/feb/22/trump-can-save-his -presidency-with-a-great-deal-to-save-the-climate.

32. Laurie Goodstein, "Evolution Slate Outpolls Rivals," *New York Times*, November 9, 2005, https://www.nytimes.com/2005/11/09/us/evolution-slate-outpolls-rivals.html.

33. Ellen Cranley, "These Are the 130 Members of Congress Who Have Doubted or Denied Climate Change," *Business Insider*, April 29, 2019, https://www.businessinsider .com/climate-change-and-republicans-congress-global-warming-2019-2#kentucky-14.

34. Other Republican mayors have now joined them. Liz Enochs, "Spotted at the Climate Summit: Republican Mayors," *Bloomberg News*, September 19, 2018, https:// www.bloomberg.com/news/articles/2018-09-19/the-republican-mayors-who-have -broken-ranks-on-climate.

35. Christopher Wolsko et al., "Red, White, and Blue Enough to Be Green: Effects of Moral Framing on Climate Change Attitudes and Conservaation Behaviors," *Journal of Experimental Social Psychology* 65 (2016), https://www.sciencedirect.com/science /article/abs/pii/S0022103116301056.

36. Dana Nuccitelli, "Trump Thinks Scientists Are Split on Climate Change. So Do Most Americans," *Guardian*, October 22, 2018, https://www.theguardian.com

/environment/climate-consensus-97-per-cent/2018/oct/22/trump-thinks-scientists -are-split-on-climate-change-so-do-most-americans; Dana Nuccitelli, "Research Shows That Facts Can Still Change Conservatives' Minds," *Guardian*, December 14, 2017, https://www.theguardian.com/environment/climate-consensus-97-per-cent/2017/dec /14/research-shows-that-certain-facts-can-still-change-conservatives-minds; Sander van der Linden et al., "Scientific Agreement Can Neutralize Politicization of Facts," *Nature Human Behaviour* 2 (January 2018), https://www.nature.com/articles/s41562 -017-0259-2; Sander van der Linden et al., "Gateway Illusion or Cultural Cognition Confusion," *Journal of Science Communication* 16, no. 5 (2017), https://jcom.sissa.it /archive/16/05/JCOM_1605_2017_A04.

37. Umair Irfan, "Report: We Have Just 12 Years to Limit Devastating Global Warming," *Vox*, October 8, 2018, https://www.vox.com/2018/10/8/17948832/climate-change -global-warming-un-ipcc-report.

38. Matt McGrath, "Climate Change: 12 Years to Save the Planet? Make That 18 Months," *BBC News*, July 24, 2019, https://www.bbc.com/news/science-environment -48964736.

39. Sarah Finnie Robinson, "How Do Americans Think about Global Warming?" Boston University: Institute for Sustainable Energy, August 9, 2018, https://www.bu .edu/ise/2018/08/09/the-51-percent-a-climate-communications-project-to-accelerate -the-transition-to-a-zero-carbon-economy/.

Chapter 6

1. Due to the pandemic, and logistical concerns, the 2020 RNC platform is identical to the 2016 platform, which refused to acknowledge the truth of anthropogenic global warming. Zoya Teirstein, "The 2020 Republican Platform: Make America 2016 Again," *Grist*, June 17, 2020, https://grist.org/politics/the-2020-republican-platform -make-america-2016-again/.

2. See my discussion of this issue in *Respecting Truth* (64–71).

3. Matt Keeley, "Only 27% of Republicans Think Climate Change Is a 'Major Threat' to the United States," *Newsweek*, August 2, 2019, https://www.newsweek.com/republi cans-climate-change-threat-1452157

4. Cary Funk, "Republicans' Views on Evolution," Pew Research, January 3, 2014, https://www.pewresearch.org/fact-tank/2014/01/03/republican-views-on-evolution -tracking-how-its-changed/.

5. As Michael Shermer points out, despite the partisan gap, such numbers are not exactly a ringing endorsement of the idea that liberals do not *ever* deny science, even on these topics. Even though the scientific consensus on climate change is 98 percent, and 97 percent on evolution, why do 16 percent of Democrats not see climate change as a major threat, and 33 percent have doubts about evolution?

Michael Shermer, "The Liberals' War on Science," *Scientific American*, February 1, 2013, https://www.scientificamerican.com/article/the-liberals-war-on-science/.

6. Stephan Lewandowsky, Jan K. Woike, and Klaus Oberauer, "Genesis or Evolution of Gender Differences? Worldview-Based Dilemmas in the Processing of Scientific Information," *Journal of Cognition* 31, no. 1 (2020), https://www.journalofcognition.org/articles/10.5334/joc.99/.

7. This hypothesis must be squared, though, with research that shows people often gravitate toward political identities that fit their preexisting values, or even their brain chemistry. Even if the specific content of their scientific beliefs may be fungible, the cognitive and personality traits that lead them to be more conservative or liberal may not be. See Chris Mooney, *The Republican Brain: The Science of Why They Deny Science—and Reality* (Hoboken, NJ: Wiley, 2012), 111–126.

8. See Michael Shermer, "The Liberals' War on Science: How Politics Distorts Science on Both Ends of the Spectrum," *Scientific American*, February 1, 2013, https://www.scientificamerican.com/article/the-liberals-war-on-science/; Chris Mooney, "The Science of Why We Don't Believe Science," *Mother Jones*, May/June 2011, https://www.motherjones.com/politics/2011/04/denial-science-chris-mooney/; Keith Kloor, "GMO Opponents Are the Climate Skeptics of the Left," *Slate*, September 26, 2012, https://slate.com/technology/2012/09/are-gmo-foods-safe-opponents-are-skewing-the-science-to-scare-people.html. See also Jon Stewart, "An Outbreak of Liberal Idiocy," *The Daily Show*, June 2, 2014, http://www.cc.com/video-clips/g1lev1/the-daily-show-with-jon-stewart-an-outbreak-of-liberal-idiocy.

9. Shermer, "The Liberals' War on Science."

10. Michael Shermer, "Science Denial versus Science Pleasure," *Scientific American*, January 1, 2018, https://www.scientificamerican.com/article/science-denial-versus-science-pleasure/.

11. Shermer, "Science Denial versus Science Pleasure."

12. Lewandowsky, Woike, and laus Oberauer, "Genesis of Evolution of Gender Differences."

13. Stephan Lewandowsky and Klaus Oberauer, "Motivated Rejection of Science," *APS: Current Direction in Psychological Science*, August 10, 2016, https://journals.sagepub.com/doi/abs/10.1177/0963721416654436?journalCode=cdpa.

14. Shermer, "The Liberals' War on Science." Note that in his 2013 essay, Shermer uses somewhat older polling data, which shows that 19 percent of Democrats doubt that the Earth is getting warmer and 41 percent are young Earth creationists.

15. That does not mean, however, that this is not an interesting question. As Tara Haelle points out, even if there is no actual "liberal war on science," it is troubling that there is any left-wing science denial at all. As she puts it, "The issue isn't

whether the Democrats are anti-science enough to match the anti-science lunacy of Republicans. The point is that any science denialism exists on the left at all." Tara Haelle, "Democrats Have a Problem with Science Too," *Politico*, June 1, 2014, https://www.politico.com/magazine/story/2014/06/democrats-have-a-problem-with -science-too-107270

16. Mooney, "The Science of Why We Don't Believe Science"; Chris Mooney, "Diagnosing the Republican Brain," *Mother Jones*, March 30, 2012, https://www .motherjones.com/politics/2012/03/chris-mooney-republican-brain-science-denial/; Chris Mooney, "There's No Such Thing as the Liberal War on Science," *Mother Jones*, March 4, 2013, https://www.motherjones.com/politics/2013/03/theres-no-such-thing -liberal-war-science/; Chris Mooney, "If You Distrust Vaccines, You're More Likely to Think NASA Faked the Moon Landings," *Mother Jones*, October 2, 2013, https://www .motherjones.com/environment/2013/10/vaccine-denial-conspiracy-theories-gmos -climate/; Chris Mooney, "Stop Pretending that Liberals Are Just as Anti-Science as Conservatives," *Mother Jones*, September 11, 2014, https://www.motherjones.com /environment/2014/09/left-science-gmo-vaccines/.

17. Though it might mean that it would be wise to rethink the hypothesis of whether science denial can be completely explained by politics, Cf. Lilliana Mason, "Ideologues without Issues: The Polarizing Consequences of Ideological Identities," *Public Opinion Quarterly*, March 21, 2018, https://academic.oup.com/poq/article/82 /S1/866/4951269

18. Even so, as Mooney points out, there might be a difference between the extent to which examples like anti-vaxx and anti-GMO can be found among liberals, versus whether it has been institutionalized as part of the Democratic Party platform, as has been the case with climate denial for the Republicans.

19. Mooney suggests that some of the partisan difference here is actually to be found at the level of brain science. "Political conservatives seem to be very different from political liberals at the level of psychology and personality. And inevitably, this influences the way the two groups argue and process information." See Mooney, "Diagnosing the Republican Brain."

20. Mooney, "The Science of Why We Don't Believe in Science."

21. But see also Chris Mooney, "Liberals Deny Science Too" (*Washington Post*, October 28, 2014), where he offers the academic left's response to evolutionary psychology as a good example of science denial: https://www.washingtonpost.com /news/wonk/wp/2014/10/28/liberals-deny-science-too/. Mooney has apparently abandoned his earlier views that anti-vaxx specifically is a good example of left-wing science denial. See Mooney, "More Polling Data on the Politics of Vaccine Resistance," *Discover Magazine*, April 27, 2011, https://www.discovermagazine.com /the-sciences/more-polling-data-on-the-politics-of-vaccine-resistance; Mooney, "The Biggest Myth about Vaccine Deniers: That They're All a Bunch of Hippie Liberals,"

Washington Post, January 26, 2015, https://www.washingtonpost.com/news/energy
-environment/wp/2015/01/26/the-biggest-myth-about-vaccine-deniers-that-theyre
-all-a-bunch-of-hippie-liberals/.

22. Even if not, we would still have to contend with the fact that so many anti-
vaxxers turn out to be liberal.

23. Jamelle Bouie, "Anti-Science Views Are a Bipartisan Problem," *Slate*, February
4, 2015, https://slate.com/news-and-politics/2015/02/conservatives-and-liberals-hold
-anti-science-views-anti-vaxxers-are-a-bipartisan-problem.html.

24. Ross Pomeroy, "Where Conservatives and Liberals Stand on Science," *Real Clear
Science*, June 30, 2015, https://www.realclearscience.com/journal_club/2015/07/01
/where_conservatives__liberals_stand_on_science.html.

25. One should note, though, that Stephan Lewandowsky challenges the idea that
anti-vaxx is bipartisan, and instead believes that it comes mostly from the political
right. This still leaves the intriguing question of whether, to the extent that both liberals and conservatives believe in anti-vaxx, they do so for the same reasons. See Lewandowsky, "Genesis or Evolution of Gender Differences? Worldview-Based Dilemmas in
the Processing of Scientific Information," *Journal of Cognition* (2020). As to the question of whether anti-vaxx is nonpartisan, to the extent that it hasn't been politicized
at all, I wonder whether this might have changed since the COVID-19 pandemic.

26. Joan Conrow, "Anti-vaccine Movement Embraced at Extremes of Political
Spectrum, Study Finds," Cornell Allience for Science, June 14, 2018, https://alliance
forscience.cornell.edu/blog/2018/06/anti-vaccine-movement-embraced-extremes
-political-spectrumstudy-finds/. There is an interesting split here, though, on the question of whether people were anti-vaxx because they felt that vaccines were unsafe or
because they were against government mandates. Charles McCoy, "Anti-vaccination
Beliefs Don't Follow the Usual Political Polarization," *The Conversation*, August 23,
2017, https://theconversation.com/anti-vaccination-beliefs-dont-follow-the-usual
-political-polarization-81001. Here there was a partisan split. "Polls Show Emerging
Ideological Divide over Childhood Vaccinations," *The Hill*, March 14, 2019, https://
thehill.com/hilltv/what-americas-thinking/434107-polls-show-emerging-ideological
-divide-over-childhood. What is fascinating is that what started as a liberal ideology may have grown in its appeal to conservatives, but for different reasons. Arthur
Allen, "How the Anti-vaccine Movement Crept into the GOP Mainstream," *Politico*,
May 27, 2019, https://www.politico.com/story/2019/05/27/anti-vaccine-republican
-mainstream-1344955.

27. Charles McCoy, "Anti-vaccination Beliefs," *The Conversation*, August 23, 2017,
https://theconversation.com/anti-vaccination-beliefs-dont-follow-the-usual-political
-polarization-81001.

28. Mooney considers Stephan Lewandowsky's work and explores the intriguing
question of whether liberals and conservatives might be anti-vaxx for different

reasons. Perhaps the right-wingers object to government intrusion in their lives, whereas left-wingers distrust the big pharmaceutical companies. Chris Mooney, "The Biggest Myth about Vaccine Deniers," *Washington Post*, January 26, 2015, https://www.washingtonpost.com/news/energy-environment/wp/2015/01/26/the -biggest-myth-about-vaccine-deniers-that-theyre-all-a-bunch-of-hippie-liberals/.

29. It is notable that not only anti-vaxxers but also white nationalists have been showing up to anti-lockdown protest rallies. Adam Bloodworth, "What Draws the Far Right and Anti-Vaxxers to Lockdown Protests," *Huffington Post*, May 17, 2020, https://www.huffingtonpost.co.uk/entry/anti-lockdown-protests-far-right-extremist -groups_uk_5ebe761ec5b65715386cb20d.

30. Jonathan Berman, *Anti-vaxxers: How to Challenge a Misinformed Movement* (Cambridge, MA: MIT Press, 2020).

31. Roni Caryn Rabin, "What Foods Are Banned in Europe But Not Banned in the US?" *New York Times*, December 28, 2018, https://www.nytimes.com/2018/12/28 /well/eat/food-additives-banned-europe-united-states.html.

32. Keith Kloor, "GMO Opponents Are the Climate Skeptics of the Left," *Slate*, September 26, 2012, https://slate.com/technology/2012/09/are-gmo-foods-safe-opponents -are-skewing-the-science-to-scare-people.html; Pamela Ronald, "Genetically Engineer Crops—What, How and Why," *Scientific American*, August 11, 2011, https://blogs .scientificamerican.com/guest-blog/genetically-engineered-crops/; Michael Gerson, "Are You Anti-GMO? Then You're Anti-science, Too," *Washington Post*, May 3, 2018, https://www.washingtonpost.com/opinions/are-you-anti-gmo-then-youre-anti -science-too/2018/05/03/cb42c3ba-4ef4-11e8-af46-b1d6dc0d9bfe_story.html; Committee on Genetically Engineered Crops et al., *Genetically Engineered Crops: Experiences and Prospects (2016)* (Washington, DC: The National Academies Press, 2016), https:// www.nap.edu/catalog/23395/genetically-engineered-crops-experiences-and-prospects; Ross Pomeroy, "Massive Review Reveals Consensus on GMO Safety," *Real Clear Science*, September 30, 2013, https://www.realclearscience.com/blog/2013/10/massive-review -reveals-consensus-on-gmo-safety.html; Jane E. Brody, "Are G.M.O. Foods Safe?" *New York Times*, April 23, 2018, https://www.nytimes.com/2018/04/23/well/eat/are-gmo -foods-safe.html.

33. The precautionary principle says that we should be careful of jumping to conclusions, especially ones that might lead to unnecessary risk. We might think that the choice to avoid GMOs could rely heavily on this principle, but perhaps that is a luxury open only to those who live in countries where food is relatively cheap and widely available. Is it reasonable to insist that GMOs must be "proven safe" (which can never happen) when millions of people around the world suffer from starvation? Perhaps the most prudent thing is to balance risk. If there is no evidence that GMOs are unsafe, and people are dying of starvation in the meantime, isn't it worth the "risk"? The parallel to vaccines here is obvious. Within the next half century, there may be as many as nine billion people on this planet. Without improvements in

food technology, how will we feed them? See Mitch Daniels, "Avoiding GMOs Isn't Just Anti-Science. It's Immoral," *Washington Post*, December 27, 2017, https://www .washingtonpost.com/opinions/avoiding-gmos-isnt-just-anti-science-its-immoral /2017/12/27/fc773022-ea83-11e7-b698-91d4e35920a3_story.html; "The World Population Prospects: 2015 Revision," UN Department of Economic and Social Affairs, July 29, 2015, https://www.un.org/en/development/desa/publications/world-population -prospects-2015-revision.html; Mark Lynas, "Time to Call Out the Anti-GMO Conspiracy," April 29, 2013, https://www.marklynas.org/2013/04/time-to-call-out-the -anti-gmo-conspiracy-theory/.

34. In "Climate Change Conspiracy Theories," Joseph E. Uscinski, Karen Douglas, and Stephan Lewandowsky write: "Conspiracy theories about genetically modified food (GMO) usually claims that a bio-tech corporation called Monsanto is engaged in a plot to overtake the agriculture industry with poisonous food." *Oxford Research Encyclopedia of Climate Science*, September 26, 2017, https://oxfordre.com/climatescience /view/10.1093/acrefore/9780190228620.001.0001/acrefore-9780190228620-e-328.

35. For a moving and eloquent discussion of the clash between what he calls "the organic fetish" and the realities of global population growth, poverty, and starvation, see chapter 3 of Michael Specter's *Denialism* (New York: Penguin Press, 2009).

36. Note that technically speaking, genetically modified organisms can be either plants or animals. And the vocabulary of genetic modification must be made clear. Every time a farmer chooses to use one plant over another, they are engaging in artificial selection, which will affect the future gene pool. More hands-on modification occurs with breeding, which can involve more traditional practices such as grafting or the newer molecular ones, where foreign genes are transferred into the genome itself. The latter is direct genetic modification. See Sheldon Krimsky, *GMOs Decoded: A Skeptic's View of Genetically Modified Foods* (Cambridge, MA: MIT Press, 2019), xxi.

37. Shahla Wunderlich and Kelsey G. Gatto, "Consumer Perception of Genetically Modified Organisms and Sources of Information," *Advances in Nutrition*, November 10, 2015, https://www.ncbi.nlm.nih.gov/pmc/articles/PMC4642419/.

38. Though it depends on what we think of as genetically modified. Does selective breeding (artificial selection) count? How about grafting? Hybridization? Gene editing? See Keith Weller, "What You Need to Know about Genetically Modified Organisms," *IFL Science*, accessed September 1, 2020, https://www.iflscience.com /environment/myths-and-controversies-gmos-0/; Laura Parker, "The GMO Labeling Battle Is Heating Up—Here's Why," *National Geographic*, January 12, 2014, https:// www.nationalgeographic.com/news/2014/1/140111-genetically-modified-organisms -gmo-food-label-cheerios-nutrition-science/#close; Elizabeth Weise, "Q&A: What You Need to Know about Genetically Engineered Foods," *USA Today*, November 19, 2015, https://www.usatoday.com/story/news/2015/11/19/what-you-need-know-genetically -engineered-foods/76059166/.

39. Weller, "What You Need to Know."

40. Anastasia Bodnar, "The Scary Truth behind Fear of GMOs," *Biology Fortified*, February 27, 2018, https://biofortified.org/2018/02/scary-truth-gmo-fear/.

41. Shahla Wunderlilch and Kelsey A. Gatto, "Consumer Perception of Genetically Modified Organisms and Sources of Information," *Advances in Nutrition* 6, no. 6 (2015), https://www.ncbi.nlm.nih.gov/pmc/articles/PMC4642419/.

42. Michael Shermer, "Are Paleo Diets More Natural Than GMOs?" *Scientific American*, April 1, 2015, https://www.scientificamerican.com/article/are-paleo-diets-more -natural-than-gmos/

43. Weller, "What You Need to Know."

44. Weller, "What You Need to Know"; Lynas, "Time to Call Out the Anti-GMO Conspiracy." Other environmentally oriented nonprofit organizations that have questioned various aspects of GMOs include Friends of the Earth and the Union of Concerned Scientists: https://foe.org/news/2015-02-are-gmos-safe-no-consensus -in-the-science-scientists/; Doug Gurian-Sherman, "Do We Need GMOs?" Union of Concerned Scientists, November 23, 2015, https://blog.ucsusa.org/doug-gurian -sherman/do-we-need-gmos-322; Keith Kloor, "On Double Standards and the Union of Concerned Scientists," *Discovery Magazine*, August 22, 2014, https://www .discovermagazine.com/environment/on-double-standards-and-the-union-of -concerned-scientists.

45. Patricia Cohen, "Roundup Weedkiller Is Blamed for Cancers, but Farmers Say It's Not Going Away," *New York Times*, September 20, 2019, https://www.nytimes .com/2019/09/20/business/bayer-roundup.html; Hilary Brueck, "The EPA Says a Chemical in Monsanto's Weed Killer Doesn't Cause Cancer—but There's Compelling Evidence That the Agency Is Wrong," *Business Insider*, June 17, 2019, https:// www.businessinsider.com/glyphosate-cancer-dangers-roundup-epa-2019-5.

46. Many environmentalists object to some of Monsanto's business practices, such as genetically modifying their seeds not to reproduce—"sterile seeds"—so that farmers will be forced to repurchase them (along with more herbicide) every year. See Mark Lynas, *Seeds of Science: Why We Go It So Wrong on GMOs* (London: Bloomsbury Sigma, 2018), 110.

47. Weller, "What You Need to Know."

48. Lynas, "Time to Call Out the Anti-GMO Conspiracy."

49. Weller, "What You Need to Know"; Gerson, "Are You Anti-GMO?"

50. "Statement by the AAAS Board of Directors on the Labeling of Genetically Modified Foods," American Association for the Advancement of Science, https://www .aaas.org/sites/default/files/AAAS_GM_statement.pdf.

51. Though what counts as a GMO can vary from place to place. The US Department of Agriculture says that gene editing is akin to crop breeding, and so doesn't count as genetic modification. In Europe, it most certainly does count. Emma Sarappo, "The Less People Understand Science, The More Afraid of GMOs They Are," *Pacific Standard*, November 19, 2018, https://psmag.com/news/the-less-people-understand -science-the-more-afraid-of-gmos-they-are; Wunderlich and Gatto, "Consumer Perception of Genetically Modified Organisms and Sources of Information."

52. Note that the term "GMO free" is prohibited in food labeling in the US, because it is impossible to test for the low-level presence of GM ingredients, and there is no federally established minimum threshold. All that can be claimed is that a product was not made through genetic engineering processes. The ubiquitous label "Non-GMO Project Verified" is run by the Non-GMO Project—a nonprofit group—to indicate that a product contains no more than 0.9 percent GMO ingredients, which is the European threshold. Wunderlich and Gatto, "Consumer Perception of Genetically Modified Organisms and Sources of Information."

53. Roberto A. Ferdman, "Why We're So Scared of GMOs, According to Someone Who Has Studied Them Since the Start," *Washington Post*, July 6, 2015, https://www .washingtonpost.com/news/wonk/wp/2015/07/06/why-people-are-so-scared-of -gmos-according-to-someone-who-has-studied-the-fear-since-the-start/.

54. Brian Kennedy et al., "Americans Are Narrowly Divided over Health Effects of Genetically Modified Foods," Pew Research, November 19, 2018, https://www .pewresearch.org/fact-tank/2018/11/19/americans-are-narrowly-divided-over-health -effects-of-genetically-modified-foods/. That view did not change over the next two years. Cary Funk, "About Half of U.S. Adults Are Wary of Health Effects of Genetically Modified Foods, but Many Also See Advantages," Pew Research, March 18, 2020, https://www.pewresearch.org/fact-tank/2020/03/18/about-half-of-u-s-adults-are-wary -of-health-effects-of-genetically-modified-foods-but-many-also-see-advantages/.

55. Brad Plumer, "Poll: Scientists Overwhelmingly Think GMOS Are Safe to Eat. The Public Doesn't," *Vox*, January 29, 2015, https://www.vox.com/2015/1/29 /7947695/gmos-safety-poll; "Public and Scientists' Views on Science and Society," Pew Research Center, January 29, 2015, https://www.pewresearch.org/science/2015 /01/29/public-and-scientists-views-on-science-and-society/#_Chapter_3:_Attitudes. The results of two other surveys are also notable. In 2013, 54 percent said that they knew little or nothing about genetically modified foods, while 25 percent had never heard of them. By 2016, 79 percent said that GMOs were dangerous. William K. Hallman et al., "Public Perceptions of Labeling Genetically Modified Foods," Rutgers Working Paper, November 1, 2013, http://humeco.rutgers.edu/documents_PDF /news/GMlabelingperceptions.pdf; Anastasia Bodnar, "The Scary Truth behind Fear of GMOs," *Biology Fortified*, February 27, 2018, https://biofortified.org/2018/02/scary -truth-gmo-fear/.

56. Quotation from Shawn Otto, *The War on Science: Who's Waging It, Why It Matters, What We Can Do about It* (Minneapolis: Milkweed, 2016), 135.

57. Where the gap was 37 percentage points. "Public and Scientists' Views on Science and Society," Pew Research, https://www.pewresearch.org/science/2015/01/29/public-and-scientists-views-on-science-and-society/.

58. In some quarters, the assumption seems to be that anything "natural" is good, so anything unnatural must therefore be bad. Yet formaldehyde is naturally found in milk, meat, and produce, and is a known carcinogen that our bodies both manufacture and metabolize. Other hypotheses include the idea that (1) it "just makes sense" to think that genetically altered food is bad for us, or that (2) GMOs somehow offend our "moral" sensibilities and cause a sense of "disgust." Sarappo, "The Less People Understand Science the More Afraid of GMOs They Are"; Roberto Ferdman, "Why We're So Scared of GMOs," *Washington Post*, July 6, 2015, https://www.washingtonpost.com/news/wonk/wp/2015/07/06/why-people-are-so-scared-of-gmos-according-to-someone-who-has-studied-the-fear-since-the-start/; Shermer, "Are Paleo Diets More Natural Than GMOs?"; Jesse Singal, "Why Many GMO Opponents Will Never Be Convinced Otherwise," *The Cut*, May 24, 2016, https://www.thecut.com/2016/05/why-many-gmo-opponents-will-never-be-convinced-otherwise.html; Stefaan Blancke, "Why People Oppose GMOs Even Though Science Says They Are Safe," *Scientific American*, August 18, 2015, https://www.scientificamerican.com/article/why-people-oppose-gmos-even-though-science-says-they-are-safe.

59. Sarappo, "The Less People Understand Science the More Afraid of GMOs They Are"; John Timmer, "On GMO Safety, the Fiercest Opponents Understand the Least," *Ars Technica*, January 15, 2019, https://arstechnica.com/science/2019/01/on-gmo-safety-the-fiercest-opponents-understand-the-least/.

60. In another result from the same study, 33 percent of respondents thought that non-GMO tomatoes contained no genes at all. Ilya Somin, "New Study Confirms That 80 Percent of Americans Support Labeling of Foods Containing DNA," *Washington Post*, March 27, 2016, https://www.washingtonpost.com/news/volokh-conspiracy/wp/2016/05/27/new-study-confirms-that-80-percent-of-americans-support-mandatory-labeling-of-foods-containing-dna/.

61. Sarappo, "The Less People Understand Science the More Afraid of GMOs They Are."

62. Gilles-Éric Séralini et al., "RETRACTED: Long-Term Toxicity of a Roundup Herbicide and Roundup-Tolerant Modified Maize," *Food and Chemical Toxicology* 50, no. 11 (November 2012): 4221–4231.

63. Weller, "What You Need To Know." Another problem with Séralini's work is that he made reporters sign nondisclosure agreements and banned them from seeking

outside opinions from other scientists before they wrote about his results. This is quite nonstandard in science. Kloor, "GMO Opponents Are the Climate Skeptics of the Left." One should note, though, that just as with Andrew Wakefield's discredited study on vaccines and autism, Séralini's study is still out there (after having been republished in another venue) and is often cited as one of the studies that provides scientific evidence against the safety of GMOs. Lynas, *Seeds of Science*, 236–237.

64. Lynas, "Time to Call Out the Anti-GMO Conspiracy."

65. Lynas, *Seeds of Science*. For what it's worth, note that Bill Nye (The Science Guy) has also done an about face on GMOs in recent years. Ross Pomeroy, "Why Bill Nye Changed His Mind on GMOs," *Real Clear Science*, October 16, 2016, https://www.realclearscience.com/blog/2016/10/why_bill_nye_changed_his_mind_on_gmos_109763.html.

66. Lynas, *Seeds of Science*, 44.

67. "Mark Lynas on His Conversion to Supporting GMOs—Oxford Lecture on Farming," YouTube, January 22, 2013, https://www.youtube.com/watch?v=vf86QYf4Suo.

68. Lynas, *Seeds of Science*, 251–252.

69. Jonathan Haidt, *The Righteous Mind: Why Good People Are Divided by Politics and Religion* (New York: Vintage, 2012), 59.

70. Lynas, *Seeds of Science*, 248.

71. "It was ironic that at just the moment the scientific community was beginning to realise that many experts' initial fears about recombinant DNA had probably been overblown, the environmental movement was solidifying its position into one of implacable opposition." Lynas, *Seeds of Science*, 172.

72. Quotation from an EarthFirst! activist, as quoted in Lynas, *Seeds of Science*, 183.

73. Lynas, *Seeds of Science*, 191.

74. Lynas, *Seeds of Science*, 188.

75. One hears echoes here of the cigarette companies' and oil companies' earlier denialist campaigns, though in this case the anti-GMO campaign was decidedly anti-corporate.

76. Lynas, *Seeds of Science*, 237.

77. Lynas, *Seeds of Science*, 211.

78. Lynas, *Seeds of Science*, 189.

79. Lynas, *Seeds of Science*, 257.

80. Lynas, *Seeds of Science*, 266–269.

81. Lynas, "Time to Call Out the Anti-GMO Conspiracy."

82. Lynas, "Time to Call Out the Anti-GMO Conspiracy." An unavoidable parallel here is former South African President Thabo Mbeki's insistence that AIDS treatments were part of a Western plot, which led to over 300,000 deaths. See Michael Specter, *Denialism: How Irrational Thinking Hinders Scientific Progress, Harms the Planet, and Threatens Our Lives* (New York: Penguin, 2009), 184, and my discussion in chapter 8. Henri E. Cauvin, "Zambian Leader Defends Ban on Genetically Altered Foods," *New York Times*, September 4, 2002, https://www.nytimes.com/2002/09/04/world/zambian-leader-defends-ban-on-genetically-altered-foods.html.

83. There are some legitimate areas of concern over GMOs, such as the evolution of superweeds in response to herbicides, pollen contamination (which could lead to less biodiversity), the potential for genetically based allergens to be introduced into new foods, pesticides lingering in the soil long after harvest, increased potential for antibiotic resistance, and so on. Scientists are working on all of these. These problems notwithstanding, no scientific study has ever shown that GMO foods are unsafe to eat. There is potential risk with any new technology; as we've seen with vaccines, the risk is low but not zero. The bottom line, then, is whether we are willing to balance these risks based on scientific evidence or resort to denialist suspicions.

84. See Stephan Lewandowsky's work on the link between conspiracy theories and science denial, cited in Mark Lynas, "Time to Call Out the Anti-GMO Conspiracy"; https://en.wikipedia.org/wiki/GMO_conspiracy_theories; Ross Pomeroy, "Why Bill Nye Changed His Mind."

85. Greenpeace, "Twenty Years of Failure: Why GM Crops Have Failed to Deliver on Their Promises," November 2015, https://storage.googleapis.com/planet4-international-stateless/2015/11/7cc5259f-twenty-years-of-failure.pdf; Lynas, *Seeds of Science*, 264.

86. Joseph E. Uscinki et al., "Climate Change Conspiracy Theories," in *Climate Science*, https://oxfordre.com/climatescience/view/10.1093/acrefore/9780190228620.001.0001/acrefore-9780190228620-e-328.

87. Lynas, "Time to Call Out the Anti-GMO Conspiracy."

88. Ivan Oransky, "Controversial Seralini GMO-Rats Paper to Be Retracted," *Retraction Watch*, November 28, 2013, https://retractionwatch.com/2013/11/28/controversial-seralini-gmo-rats-paper-to-be-retracted/.

89. Steven Novella, "Golden Rice Finally Released in Bangladesh," *Neurologica* (blog), March 8, 2019, https://theness.com/neurologicablog/index.php/golden-rice-finally-released-in-bangladesh/.

90. Sheldon Krimsky, *GMOs Decoded: A Skeptic's View of Genetically Modified Foods* (Cambridge, MA: MIT Press, 2019).

91. Krimsky, *GMOs Decoded*, xviii.

92. For instance, some naturally grown potatoes are toxic, due to high levels of glycoalkaloids. Krimsky, *GMOs Decoded*, 73, 107.

93. Krimsky, *GMOs Decoded*, 74–75.

94. Kriimsky, *GMOs Decoded*, 75.

95. Krimsky, *GMOs Decoded*, 79.

96. Note though that 12 percent dissent (on GMOs) is not the same as 1 percent (on climate change). The question of consensus is not merely one of a public knowledge gap. Indeed one might wonder: if 12 percent of scientists doubt the safety of GMOs, is there really a consensus? Yet even here, skepticism is allowed about the narrow remaining areas of debate, but denialism is not.

97. H. J. Mai, "U.N. Warns Number of People Starving to Death Could Double Amid Pandemic," *NPR*, May 5, 2020, https://www.npr.org/sections/coronavirus-live-updates/2020/05/05/850470436/u-n-warns-number-of-people-starving-to-death-could-double-amid-pandemic.

98. Krimsky, *GMOs Decoded*, 124, 149.

99. Krimsky, *GMOs Decoded*, 87.

100. See here my argument in *The Scientific Attitude*, 29–34.

101. But note that consensus does not require 100 percent agreement. Lynas, *Seeds of Science*, 260.

102. Krimsky, *GMOs Decoded*, 104.

103. Krimsky, *GMOs Decoded*, 115. Note that similar types of comparative studies on whether thimerosal (in vaccines) caused autism—given that thimerosal had been banned in Europe before it was in the US—are taken to be definitive evidence to debunk the claim that vaccines cause autism.

Chapter 7

1. H. Claire Brown and Joe Fassler, "Whole Foods Quietly Pauses Its GMO Labeling Requirements," *The Counter*, May 21, 2018, https://thecounter.org/whole-foods-gmo-labeling-requirements/.

2. "GMO Labeling," Whole Foods Market, accessed September 1, 2020, https://www.wholefoodsmarket.com/quality-standards/gmo-labeling.

3. Adam Campbell-Schmitt, "Whole Foods Pauses GMO Labeling Deadline for Suppliers," *Food and Wine*, May 22, 2018, https://www.foodandwine.com/news/whole-foods-gmo-labeling-policy.

4. Michael Schulson, "Whole Foods: America's Temple of Pseudoscience," *Daily Beast*, May 20, 2019, https://www.thedailybeast.com/whole-foods-americas-temple-of-pseudoscience.

5. Michael Shermer, "The Liberals' War on Science," *Scientific American*, Feburary 1, 2013, https://www.scientificamerican.com/article/the-liberals-war-on-science/.

6. When I called back the next week with a few follow-up questions, she said this was a tough issue, but why not try to get the nutrients into the food some other way, without having to support the GMO industry?

7. I have to admit, I was intrigued, so I did some research and found that about 5 percent of wheat farmers use Roundup as a desiccant to kill the wheat stalks just before harvesting, which makes it dry and easier to harvest. On the question of whether there is a safety risk to this practice, see David Mikkelson and Alex Kasprak, "The Real Reason Wheat Is Toxic," *Snopes*, December 25, 2014 (updated July 26, 2017), https://www.snopes.com/fact-check/wheat-toxic/.

8. During that second call, she clarified that her worries about food safety and the environment were linked. If we poison the soil, aren't we also potentially harming the future food supply? She also said that the scientific consensus on safety for GMO foods didn't do much to measure these kinds of downstream effects.

9. I'd read that planting trees was one of the best ways to mitigate the effects of climate change, and I let him calculate how many. Mark Tutton, "The Most Effective Way to Tackle Climate Change? Plant 1 Trillion Trees," *CNN*, April 17, 2019, https://www.cnn.com/2019/04/17/world/trillion-trees-climate-change-intl-scn/index.html.

10. It is in fact the plot of the movie *I Am Legend*, which Ted hadn't seen.

11. And of course relying on bogus or made-up evidence for your skepticism—or having no evidence whatsoever as the basis for your concerns—constitutes denialism too.

12. Stephan Lewandowsky, Jan K. Woke, and Klaus Oberauer, "Genesis or Evolution of Gender Differences," *Journal of Cognition* 31, no. 1 (2020): 1–25, https://www.journalofcognition.org/articles/10.5334/joc.99/.

13. Stephan Lewandowsky and Klaus Oberauer, "Motivated Rejection of Science," *Current Directions in Psychological Science* 25, no. 4 (2016): 217–222.

14. Lawrence Hamilton, "Conservative and Liberal Views of Science," *Carsey Research Regional Issue Brief* 45 (Summer 2015), https://scholars.unh.edu/cgi/viewcontent.cgi?article=1251&context=carsey.

15. Hamilton found that "Liberal-conservative gaps on these questions ranged from 55 points (climate change) to 24 points (nuclear power), but always in the same

direction." That is, there was no area where liberals had less trust in scientists than conservatives did.

16. Brian Kennedy and Cary Funk, "Many Americans Are Skeptical about Scientific Research on Climate and GM Foods," Pew Research, December 5, 2016, https://www.pewresearch.org/fact-tank/2016/12/05/many-americans-are-skeptical-about-scientific-research-on-climate-and-gm-foods/.

17. But of course then anti-GMO would be a case of conservative science denial as well. So it really depends how one wants to frame the question. Technically, given the bipartisan polling results, neither anti-vaxx nor anti-GMO look like a good candidate for the mantle of liberal science denial. But that does not mean there isn't a problem with many liberals' views of some scientific topics. One intriguing question here, though, is whether both of these topics started as liberal, then became bipartisan. See Langer (2001) cited in Joseph E. Uscinski, Karen Douglas, and Stephan Lewandowsky, "Climate Change Conspiracy Theories," *Climate Science*, September 26, 2017, https://oxfordre.com/climatescience/view/10.1093/acrefore/9780190228620.001.0001/acrefore-9780190228620-e-328.

18. Stephan Lewandowsky, G. E. Gignac, and K. Oberauer, "The Role of Conspiracist Ideation and Worldviews in Predicting Rejection of Science," *PLoS ONE* 10, no. 8 (2015), https://journals.plos.org/plosone/article?id=10.1371/journal.pone.0075637.

19. Lewandowsky, Gignac, and Oberauer, "The Role of Conspiracist Ideation and Worldviews." To be clear, we could measure a subject's political ideology by whether they agreed or disagreed with a conservative worldview. The same is true for free market ideology. For an explanation of the predictive power of the different valences of these worldviews on anti-vaxx, see Lewandowsky et al.'s "Conspiracist Ideation." But the point for GMOs is that there was no correlation at all with one's worldview, one way or the other.

20. This is probably a better way to analyze the question than merely looking at the numbers of partisans who say that they agree or disagree with any particular scientific consensus, as Hamilton does. For even if there were more liberals than conservatives who were GMO deniers, would this in and of itself show that anti-GMO was an example of liberal science denial? Probably not, for as Lewandowsky's work shows, one must also account for the ideology behind the partisan label.

21. Lewandowsky, Gignac, and Oberauer, "The Role of Conspiracist Ideation and Worldviews."

22. Charles McCoy, "Anti-vaccination Beliefs Don't Follow the Usual Political Polarization," *The Conversation*, August 23, 2017, https://theconversation.com/anti-vaccination-beliefs-dont-follow-the-usual-political-polarization-81001; Joan Conrow, "Anti-vaccine Movement Embraced at Extremes of Political Spectrum, Study Finds," Cornell Alliance for Science, June 14, 2018, https://allianceforscience.cornell.edu

/blog/2018/06/anti-vaccine-movement-embraced-extremes-political-spectrumstudy
-finds/; Matthew Sheffield, "Polls Show Emerging Ideological Divide Over Childhood
Vaccinations," *The Hill*, March 14, 2019, https://thehill.com/hilltv/what-americas
-thinking/434107-polls-show-emerging-ideological-divide-over-childhood.

23. Recall that there is reason for thinking this might be true, given that with anti-
vaxx there was a partisan split between anti-government and anti-big-pharma ideology. With GMOs, it seems to be anti-government versus anti-corporate. That seems
pretty close. But see Dan Kahan, "We Aren't Polarized on GM Foods—No Matter
What the Result in Washington State," Cultural Cognition Project, November 5,
2013, http://www.culturalcognition.net/blog/2013/11/5/we-arent-polarized-on-gm
-foods-no-matter-what-the-result-in.html; and also Dan Kahan, "Trust in Science
& Perceptions of GM Food Risks—Does the GSS Have Something to Say on This?"
Cultural Cognition Project, March 16, 2017, http://www.culturalcognition.net/blog
/2017/3/16/trust-in-science-perceptions-of-gm-food-risks-does-the-gss-h.html.

24. Lewandowsky and Oberauer, "Motivated Rejection of Science."

25. See here Chris Mooney's discussion of evolutionary psychology in "Liberals Deny
Science, Too," *Washington Post*, October 28, 2014, https://www.washingtonpost.com
/news/wonk/wp/2014/10/28/liberals-deny-science-too/. See also Michael Shermer's
"Science Denial versus Science Pleasure," *Scientific American*, January 1, 2018, https://
www.scientificamerican.com/article/science-denial-versus-science-pleasure/ and Sher-
man, "The Liberals' War on Science."

26. Uscinki et al., "Climate Change Conspiracy Theories."

27. But does this mean therefore that all science denial comes from the right? No.
To say that one hasn't found enough evidential support to conclude that GMO
denial comes from the left does not automatically mean that it comes from the
right. In fact the *very same evidence* Lewandowsky cites to show that GMO denial
does not come from the left can be used to show that it does not come from the
right. "No correlation" swings both ways.

28. Lewandowsky et al., "The Role of Conspiracist Ideation."

29. An important 2017 study by Anthony Washburn and Linda Skitka examined
precisely this question, and confirmed that both liberals and conservatives were
equally likely to use motivated reasoning, when a scientific result conflicted with
their existing beliefs. Anthony N. Washburn and Linda J. Skitka, "Science Denial
Across the Political Divide: Liberals and Conservatives and Similarly Motivated to
Deny Attitude-Inconsistent Science," *Social Psychology and Personality Science* 9, no. 9
(2018), https://lskitka.people.uic.edu/WashburnSkitka2017_SPPS.pdf.

30. Tara Haelle, "Democrats Have a Problem with Science, Too," *Politico*, June 1 2014,
https://www.politico.com/magazine/story/2014/06/democrats-have-a-problem-with
-science-too-107270. See also Eric Armstrong, "Are Democrats the Party of Science?

Not Really," *New Republic*, January 10, 2017, https://newrepublic.com/article/139700
/democrats-party-science-not-really.

31. Donnelle Eller, "Anti-GMO Articles Tied to Russian Sites, ISU Research Shows," *Des Moines Register*, February 25, 2018, https://www.desmoinesregister.com/story/money /agriculture/2018/02/25/russia-seeks-influence-usa-opinion-gmos-iowa-state-research /308338002/; Justin Cremer, "Russia Uses 'Information Warfare' to Portray GMOs Negatively," Cornell Alliance for Science, February 28, 2018, https://allianceforscience .cornell.edu/blog/2018/02/russia-uses-information-warfare-portray-gmos-negatively/. Indeed, according to the *New York Times*, the Russian government has been spreading science denial propaganda since the AIDS crisis in the 1980s, up through Ebola, until today with numerous conspiracies about the cause of the coronavirus. William J. Broad, "Putin's Long War Against American Science," *New York Times*, April 13, 2020, https://www.nytimes.com/2020/04/13/science/putin-russia-disinformation-health -coronavirus.html; Julian E. Barnes and David E. Sanger, "Russian Intelligence Agencies Push Disinformation on Pandemic," *New York Times*, July 28, 2020, https:// www.nytimes.com/2020/07/28/us/politics/russia-disinformation-coronavirus.html. For more information and citations about Russian propaganda efforts, see chapter 8.

Chapter 8

1. Sarah Boseley, "Mbeki AIDS Denial 'Caused 300,000 Deaths,'" *Guardian*, November 26, 2008, https://www.theguardian.com/world/2008/nov/26/aids-south-africa.

2. Epidemiologists have estimated that approximately 90 percent of American deaths from coronavirus were due to the Trump administration's delay between March 2 and March 16. Eugene Jarecki, "Trump's Covid-19 Inaction Killed Americans. Here's a Counter that Shows How Many," *Washington Post*, May 6, 2020, https://www .washingtonpost.com/outlook/2020/05/06/trump-covid-death-counter/.

3. Joseph Uscinski et al., "Why Do People Believe COVID-19 Conspiracy Theories?," *Misinformation Review*, April 28, 2020, https://misinforeview.hks.harvard.edu/article /why-do-people-believe-covid-19-conspiracy-theories/.

4. Stephan Lewandowsky and John Cook, "Coronavirus Conspiracy Theories Are Dangerous—Here's How to Stop Them Spreading," *The Conversation*, April 20, 2020, https://theconversation.com/coronavirus-conspiracy-theories-are-dangerous-heres -how-to-stop-them-spreading-136564; Adam Satariano and Davey Alba, "Burning Cell Towers, Out of Baseless Fear They Spread the Virus," *New York Times*, April 10, 2020, https://www.nytimes.com/2020/04/10/technology/coronavirus-5g-uk.html; Travis M. Andrews, "Why Dangerous Conspiracy Theories about the Virus Spread So Fast—and How They Can Be Stopped," *Washington Post*, May 1, 2020, https://www .washingtonpost.com/technology/2020/05/01/5g-conspiracy-theory-coronavirus -misinformation/.

5. Matthew Rozsa, "We Asked Experts to Respond to the Most Common COVID-19 Conspiracy Theories and Misinformation," *Salon*, July 18, 2020, https://www.salon .com/2020/07/18/we-asked-experts-to-respond-to-the-most-common-covid-19 -conspiracy-theories/; Quassim Cassam, "Covid Conspiracies," *ABC Saturday Extra*, May 16, 2020, https://www.abc.net.au/radionational/programs/saturdayextra/covid -conspiracies/12252406.

6. William J. Broad and Dan Levin, "Trump Muses about Light as Remedy, but Also Disinfectant, Which Is Dangerous," *New York Times*, April 24, 2020, https://www .nytimes.com/2020/04/24/health/sunlight-coronavirus-trump.html.

7. Mayla Gabriela Silva Borba et al., "Chloroquine Diphosphate in Two Different Dosages As Adjunctive Therapy of Hospitalized Patients with Severe Respiratory Syndrome in the Context of Coronavirus (SARS-CoV-2) infection: Preliminary Safety Results of a Randomized, Double-Blinded, Phase IIb Clinical Trial (CloroCovid-19 Study)," *medRxiv* (preprint), April 7, 2020, https://www.medrxiv.org/content/10.1101 /2020.04.07.20056424v2; Christian Funke-Brentano et al., "Retraction and Repub- lication: Cardiac Toxicity of Hydroxychloroquine in COVID-19," *Lancet*, July 9, 2020, https://www.ncbi.nlm.nih.gov/pmc/articles/PMC7347305/; Katie Thomas and Knvul Sheikh, "Small Chloroquine Study Halted over Risk of Fatal Heart Complica- tions," *New York Times*, April 12, 2020, https://www.nytimes.com/2020/04/12/health /chloroquine-coronavirus-trump.html; Elyse Samuels and Meg Kelly, "How False Hope Spread about Hydroxychloroquine to Treat COVID-19—and the Consequences That Followed," *Washington Post*, April 13, 2020, https://www.washingtonpost .com/politics/2020/04/13/how-false-hope-spread-about-hydroxychloroquine -its-consequences/; Paul Farhi and Elahe Izadi, "Fox News Goes Mum on the COVID-19 Drug They Spent Weeks Promoting," *Washington Post*, April 23, 2020, https://www.washingtonpost.com/lifestyle/media/fox-news-hosts-go-mum-on -hydroxychloroquine-the-covid-19-drug-they-spent-weeks-promoting/2020/04/22 /eeaf90c2-84ac-11ea-ae26-989cfce1c7c7_story.html.

8. Dickens Olewe, "Stella Immanuel—the Doctor behind Unproven Coronavirus Cure Claim," *BBC News*, July 29, 2020, https://www.bbc.com/news/world-africa-53579773.

9. Margaret Sullivan, "This Was the Week America Lost the War on Misinformation," *Washington Post*, July 30, 2020, https://www.washingtonpost.com/lifestyle/media /this-was-the-week-america-lost-the-war-on-misinformation/2020/07/30/d8359e2e -d257-11ea-9038-af089b63ac21_story.html.

10. Stephen Collinson, "Trump Seeks a 'Miracle' as Virus Fears Mount," *CNN*, Febru- ary 28, 2020, https://www.cnn.com/2020/02/28/politics/donald-trump-coronavirus -miracle-stock-markets/index.html.

11. Consider this analogy: if we were catching more fish only because we were put- ting out twice as many nets, the number of fish in each net wouldn't be going up

even if the total catch was. But that's not what is happening. Each net has more fish *and* we are putting out more nets.

12. Right-wing media has played a key role in disseminating misinformation about the coronavirus. According to one study out of the Harvard Kennedy School of Government, Fox News viewers were much more likely to downplay the threat of coronavirus, because the hosts of popular Fox News shows were doing that. Margaret Sullivan, "The Data Is In: Fox News May Have Kept Millions from Taking the Coronavirus Threat Seriously," *Washington Post*, June 28, 2020, https://www.washingtonpost.com/lifestyle /media/the-data-is-in-fox-news-may-have-kept-millions-from-taking-the-coronavirus -threat-seriously/2020/06/26/60d88aa2-b7c3-11ea-a8da-693df3d7674a_story.html. An even more fine-grained analysis has revealed that there was a correlation between specific Fox News programming and the prevalence of coronavirus cases and deaths. Zack Beauchamp, "A Disturbing New Study Suggests Sean Hannity's Show Helped Spread the Coronavirus," *Vox*, April 22, 2020, https://www.vox.com/policy-and-politics/2020 /4/22/21229360/coronavirus-covid-19-fox-news-sean-hannity-misinformation-death.

13. Dan Diamond and Nahal Toosi, "Trump Team Failed to Follow NSC's Pandemic Playbook," *Politico*, March 25, 2020, https://www.politico.com/news/2020/03 /25/trump-coronavirus-national-security-council-149285.

14. Sharon LaFraniere et al., "Scientists Worry About Political Influence over Coronavirus Vaccine Project," *New York Times*, August 2, 2020, https://www.nytimes.com /2020/08/02/us/politics/coronavirus-vaccine.html

15. Nicholas Bogel-Burroughs, "Antivaccination Activists Are Growing Force at Virus Protests," *New York Times*, May 2, 2020, https://www.nytimes.com/2020/05/02/us /anti-vaxxers-coronavirus-protests.html.

16. Liz Szabo, "The Anti-vaccine and Anti-lockdown Movements Are Converging, Refusing to Be 'Enslaved,'" *Los Angeles Times*, April 24, 2020, https://www.latimes .com/california/story/2020-04-24/anti-vaccine-activists-latch-onto-coronavirus-to -bolster-their-movement.

17. Emma Reynolds, "Some Anti-vaxxers Are Changing Their Minds because of the Coronavirus Pandemic," *CNN*, April 20, 2020, https://www.cnn.com/2020/04/20 /health/anti-vaxxers-coronavirus-intl/index.html; Jon Henley, "Coronavirus Causing Some Anti-vaxxers to Waver, Experts Say," *Guardian*, April 21, 2020, https:// www.theguardian.com/world/2020/apr/21/anti-vaccination-community-divided -how-respond-to-coronavirus-pandemic; Victoria Waldersee, "Could the New Coronavirus Weaken 'Anti-vaxxers'?" Reuters, April 11, 2020, https://www.reuters .com/article/us-health-coronavirus-antivax/could-the-new-coronavirus-weaken-anti -vaxxers-idUSKCN21T089.

18. Andrew E. Kramer, "Russia Sets Mass Vaccinations for October After Shortened Trial," *New York Times*, August 2, 2020, https://www.nytimes.com/2020/08/02/world /europe/russia-trials-vaccine-October.html.

19. Lauren Neergaard and Hananah Fingerhut, "AP-NORC Poll: Half of Americans Would Get a COVID-19 Vaccine," *AP News*, May 27, 2020, https://apnews.com/dacd c8bc428dd4df6511bfa259cfec44.

20. Steven Sparks and Gary Langer, "27% Unlikely to Be Vaccinated against the Coronavirus; Republicans, Conservatives Especially: POLL," *ABC News*, June 2, 2020, https://abcnews.go.com/Politics/27-vaccinated-coronavirus-republicans-conservatives -poll/story?id=70962377.

21. Rebecca Falconer, "Fauci: Coronavirus Vaccine May Not Be Enough to Achieve Herd Immunity in the U.S.," *Axios*, June 29, 2020, https://www.axios.com /fauci-coronavirus-vaccine-herd-immunity-unlikely-023151cc-086d-400b-a416 -2f561eb9a7fa.html.

22. Manny Fernandez, "Conservatives Fuel Protests Against Coronavirus Lock-downs," *New York Times*, April 18, 2020, https://www.nytimes.com/2020/04/18/ us/texas-protests-stay-at-home.html; Jason Wilson and Robert Evans, "Revealed: Major Anti-lockdown Group's Links to America's Far Right," *Guardian*, May 8, 2020, https://www.theguardian.com/world/2020/may/08/lockdown-groups-far-right-links -coronavirus-protests-american-revolution.

23. Chuck Todd et al., "The Gender Gap between Trump and Biden Has Turned into a Gender Canyon," *NBC News*, June 8, 2020, https://www.nbcnews.com/politics/meet -the-press/gender-gap-between-trump-biden-has-turned-gender-canyon-n1227261.

24. Neil MacFarquhar, "Who's Enforcing Mask Rules? Often Retail Workers, and They're Getting Hurt," *New York Times*, May 15, 2020, https://www.nytimes.com /2020/05/15/us/coronavirus-masks-violence.html; Bill Hutchinson, "'Incomprehen-sible': Confrontations over Masks Erupt amid COVID-19 Crisis," *ABC News*, May 7, 2020, https://abcnews.go.com/US/incomprehensible-confrontations-masks-erupt -amid-covid-19-crisis/story?id=70494577.

25. Kate Yoder, "Russian Trolls Shared Some Truly Terrible Climate Change Memes," *Grist*, May 1, 2018, https://grist.org/article/russian-trolls-shared-some-truly-terrible -climate-change-memes/; Craig Timberg and Tony Romm, "These Provocative Images Show Russian Trolls Sought to Inflame Debate over Climate Change, Fracking and Dakota Pipeline," *Washington Post*, March 1, 2018, https://www.washingtonpost .com/news/the-switch/wp/2018/03/01/congress-russians-trolls-sought-to-inflame-u -s-debate-on-climate-change-fracking-and-dakota-pipeline/; Rebecca Leber and A.J. Vicens, "7 Years Before Russia Hacked the Election, Someone Did the Same Thing to Climate Scientists," *Mother Jones*, January/February 2018, https://www.motherjones .com/politics/2017/12/climategate-wikileaks-russia-trump-hacking/.

26. Carolyn Y. Johnson, "Russian Trolls and Twitter Bots Exploit Vaccine Con-troversy," *Washington Post*, August 23, 2018, https://www.washingtonpost.com /science/2018/08/23/russian-trolls-twitter-bots-exploit-vaccine-controversy/; Jessica Glenza, "Russian Trolls 'Spreading Discord' over Vaccine Safety Online," *Guardian*,

August 23, 2018, https://www.theguardian.com/society/2018/aug/23/russian-trolls-spread-vaccine-misinformation-on-twitter.

27. Donnelle Eller, "Anti-GMO Articles Tied to Russian Sites," *Des Moines Register*, Feburary 25, 2018, https://www.desmoinesregister.com/story/money/agriculture/2018/02/25/russia-seeks-influence-usa-opinion-gmos-iowa-state-research/308338002/; Justin Cremer, "Russia Uses 'Information Warfare' to Portray GMOs Negatively," Alliance for Science, February 2018, https://allianceforscience.cornell.edu/blog/2018/02/russia-uses-information-warfare-portray-gmos-negatively/.

28. Robin Emmott, "Russia Deploying Coronavirus Disinformation to Sow Panic in West, EU Document Shows," Reuters, March 18, 2020, https://www.reuters.com/article/us-health-coronavirus-disinformation/russia-deploying-coronavirus-disinformation-to-sow-panic-in-west-eu-document-says-idUSKBN21518F.

29. Allen Kim, "Nearly Half of the Twitter Accounts Discussing 'Reopening America' May Be Bots, Researchers Say," *CNN*, May 22, 2020, https://www.cnn.com/2020/05/22/tech/twitter-bots-trnd/index.html.

30. Eric Tucker, "US Officials: Russia behind Spread of Virus Disinformation," *AP News*, July 28, 2020, https://apnews.com/3acb089e6a333e051dbc4a465cb68ee1.

31. "The Coronavirus Gives Russia and China Another Opportunity to Spread Their Disinformation," *Washington Post*, March 29, 2020, https://www.washingtonpost.com/opinions/the-coronavirus-gives-russia-and-china-another-opportunity-to-spread-their-disinformation/2020/03/29/8423a0f8-6d4c-11ea-a3ec-70d7479d83f0_story.html; Edward Wong et al., "Chinese Agents Spread Messages That Sowed Virus Panic in U.S., Officials Say," *New York Times*, April 22, 2020, https://www.nytimes.com/2020/04/22/us/politics/coronavirus-china-disinformation.html.

32. Oliver Milman, "Revealed: Quarter of All Tweets about Climate Crisis Produced by Bots," *Guardian*, February 21, 2020, https://www.theguardian.com/technology/2020/feb/21/climate-tweets-twitter-bots-analysis; Ryan Bort, "Study: Bots Are Fueling Online Climate Denialism," *Rolling Stone*, February 21, 2020, https://www.rollingstone.com/politics/politics-news/bots-fueling-climate-science-denialism-twitter-956335/.

33. Whether or not Zuckerberg wants Facebook to be the "arbiter of truth," so many people get their news from his website that perhaps it already is. Steven Levy, "Mark Zuckerberg Is an Arbiter of Truth—Whether He Likes It or Not," *Wired*, June 5, 2020, https://www.wired.com/story/mark-zuckerberg-is-an-arbiter-of-truth-whether-he-likes-it-or-not/.

34. Tony Romm, "Facebook CEO Mark Zuckerberg Says in Interview He Fears 'Erosion of Truth' but Defends Allowing Politicians to Lie in Ads," *Washington Post*, October 17, 2019, https://www.washingtonpost.com/technology/2019/10/17/facebook-ceo-mark-zuckerberg-says-interview-he-fears-erosion-truth-defends-allowing-politicians-lie-ads/.

35. Craig Timberg and Andrew Ba Tran, "Facebook's Fact-Checkers Have Ruled Claims in Trump's Ads Are False—but No One Is Telling Facebook's Users," *Washington Post*, August 5, 2020, https://www.washingtonpost.com/technology/2020/08/05 /trump-facebook-ads-false/.

36. Jason Murdock, "Most COVID-19 Misinformation Originates on Facebook, Research Suggests," *Newsweek*, July 6, 2020, https://www.newsweek.com/facebook -covid19-coronavirus-misinformation-twitter-youtube-whatsapp-1515642.

37. Heather Kelly, "Facebook, Twitter Penalize Trump for Posts Containing Coronavirus Misinformation," *Washington Post*, August 7, 2020, https://www.washingtonpost .com/technology/2020/08/05/trump-post-removed-facebook/.

38. Alex Kantrowitz, "Facebook Is Taking Down Posts That Cause Imminent Harm—but Not Posts That Cause Inevitable Harm," *BuzzFeed News*, May 3, 2020, https://www.buzzfeednews.com/article/alexkantrowitz/facebook-coronavirus -misinformation-takedowns.

39. "Twitter to Label Misinformation about Coronavirus amid Flood of False Claims and Conspiracy Theories," *CBS News*, May 13, 2020, https://www.cbsnews.com /news/twitter-misinformation-disputed-tweets-claims-coronavirus/.

40. Craig Timberg et al., "Tech Firms Take a Hard Line against Coronavirus Myths. But What about Other Types of Misinformation?" *Washington Post*, February 28, 2020, https://www.washingtonpost.com/technology/2020/02/28/facebook-twitter-amazon -misinformation-coronavirus/.

41. Michael Segalov, "The Parallels between Corornavirus and Climate Crisis Are Obvious," *Guardian*, May 4, 2020, https://www.theguardian.com/environment/2020 /may/04/parallels-climate-coronavirus-obvious-emily-atkin-pandemic; Beth Gardner, "Coronavirus Holds Key Lessons on How to Fight Climate Change," *Yale Environment 360*, March 23, 2020, https://e360.yale.edu/features/coronavirus-holds-key-lessons-on -how-to-fight-climate-change.

42. Bess Levin, "Texas Lt. Governor: Old People Should Volunteer to Die to Save the Economy," *Washington Post*, March 24, 2020, https://www.washingtonpost.com /sports/2020/04/18/sally-jenkins-trump-coronavirus-testing-economy/ https://www .vanityfair.com/news/2020/03/dan-patrick-coronavirus-grandparents.

43. John Kerry makes a different argument, which is that fighting climate change does not require us to shut down the economy, but instead would create the jobs and infrastructure to develop a renewed economy around the issues of clean energy. Rachel Koning Beals, "COVID-19 and Climate Change: 'The Parallels Are Screaming at Us,' Says John Kerry," *Market Watch*, April 22, 2020, https://www.marketwatch .com/story/covid-19-and-climate-change-the-parallels-are-screaming-at-us-says-john -kerry-2020-04-22.

44. Paul Krugman, "COVID-19 Brings Out All the Usual Zombies," *New York Times*, March 28, 2020, https://www.nytimes.com/2020/03/28/opinion/coronavirus-trump -response.html.

45. For a more detailed look at how the parallels play out around these five stages, see Lewandowsky and Cook's use of an image from Yale Climate Connections: "Coronavirus Conspiracy Theories Are Dangerous," *The Conversation*, April 20, 2020, https://theconversation.com/coronavirus-conspiracy-theories-are-dangerous-heres -how-to-stop-them-spreading-136564.

46. Beth Gardiner, "Coronavirus Holds Key Lessons," *Yale Environment 360*, March 23, 2020, https://e360.yale.edu/features/coronavirus-holds-key-lessons-on-how-to -fight-climate-change.

47. Charlie Sykes, "Did Trump and Kushner Ignore Blue State COVID-19 Test- ing as Deaths Spiked?," *NBC News*, August 4, 2020, https://www.nbcnews.com /think/opinion/did-trump-kushner-ignore-blue-state-covid-19-testing-deaths -ncna1235707.

48. Bill Barrow et al., "Coronavirus' Spread in GOP Territory, Explained in Six Charts," *AP News*, June 30, 2020, https://apnews.com/7aa2fcf7955333834e01a7f9217c77d2.

49. Lewandowsky and Oberauer, "Motivated Rejection of Science"; Sander van der Linden, Anthony Leiserorwitz, and Edward Maibach, "Gateway Illusion or Cultural Cognition Confusion?," *Journal of Science Communication* 16, no. 5 (2017), https:// jcom.sissa.it/archive/16/05/JCOM_1605_2017_A04.

50. Dana Nuccitelli, "Research Shows That Certain Facts Can Still Change Con- servatives' Minds," *Guardian*, December 14, 2017, https://www.theguardian.com /environment/climate-consensus-97-per-cent/2017/dec/14/research-shows-that -certain-facts-can-still-change-conservatives-minds.

51. Madeleine Carlisle, "Three Weeks After Trump's Tulsa Rally, Oklahoma Reports Record High COVID-19 Numbers," *Time*, July 11, 2020, https://time.com/5865890 /oklahoma-covid-19-trump-tulsa-rally/.

52. "Coronavirus: Donald Trump Wears Face Mask for the First Time," *BBC*, July 12, 2020, https://www.bbc.com/news/world-us-canada-53378439.

53. John Wagner et al., "Herman Cain, Former Republican Presidential Hopeful, Has Died of Coronavirus, His Website Says," *Washington Post*, July 30, 2020, https://www .washingtonpost.com/politics/herman-cain-former-republican-presidential-hopeful -has-died-of-the-coronavirus-statement-on-his-website-says/2020/07/30/4ac62a10 -d273-11ea-9038-af089b63ac21_story.html.

54. Ashley Collman, "A Man Who Thought the Coronavirus Was a 'Scamdemic' Wrote a Powerful Essay Warning against Virus Deniers after He Hosted a Party and Got

His Entire Family Sick," *Business Insider*, July 28, 2020, https://www.businessinsider
.com/coronavirus-texas-conservative-thought-hoax-before-infection-2020-7.

55. Janelle Griffith, "He Thought the Coronavirus Was 'a Fake Crisis.' Then He Con-
tracted It and Changed His Mind," *NBC News*, May 18, 2020, https://www.nbcnews
.com/news/us-news/he-thought-coronavirus-was-fake-crisis-then-he-contracted-it
-n1209246.

56. Kim LaCapria, "Richard Rose Dies of COVID-19, After Repeated 'Covid Denier'
Posts," *Truth or Fiction*, July 10, 2020, https://www.truthorfiction.com/richard-rose
-dies-of-covid-19-after-repeated-covid-denier-posts/.

57. Kristin Urquiza, "Governor, My Father's Death Is on Your Hands," *Washington
Post*, July 27, 2020, https://www.washingtonpost.com/outlook/governor-my-fathers
-death-is-on-your-hands/2020/07/26/55a43bec-cd15-11ea-bc6a-6841b28d9093_
story.html.

58. Charlie Warzel, "How to Actually Talk to Anti-Maskers," *New York Times*, July
22, 2020, https://www.nytimes.com/2020/07/22/opinion/coronavirus-health-experts
.html.

59. Warzel, "How to Actually Talk to Anti-Maskers."

60. He cites a June 2020, *New York Times*/Siena College survey that showed that 90
percent of Democrats, and 75 percemt of Republicans, said that they trusted medical
scientists to provide reliable information about COVID-19.

61. Warzel, "How to Actually Talk to Anti-Maskers."

62. See my book *The Scientific Attitude*.

63. Here, too, admitting uncertainty and showing a bit of humility may be effective.
Mark Honigsbaum, "Anti-vaxxers: Admitting That Vaccinology Is an Imperfect Science May Be a Better Way to Defeat Sceptics," *The Conversation*, February 15, 2019,
https://theconversation.com/anti-vaxxers-admitting-that-vaccinology-is-an-imperfect
-science-may-be-a-better-way-to-defeat-sceptics-111794?utm_medium=Social&utm_
source=Twitter#Echobox=1550235443.

64. Charlie Warzel, "How to Actually Talk to Anti-Maskers."

65. For an excellent short piece that offers practical tips on how to have effective
conversations with those who believe conspiracy theories about scientific topics, see
Tanya Basu, "How to Talk to Conspiracy Theorists—and Still Be Kind," *MIT Technology Review*, July 15, 2020, https://www.technologyreview.com/2020/07/15/1004950
/how-to-talk-to-conspiracy-theorists-and-still-be-kind/.

66. See note 35 in chapter 3 of this book for citations to some others who have
discussed this idea.

Bibliography

Appiah, Kwame Anthony. "People Don't Vote for What They Want: They Vote for Who They Are." *Washington Post*, August 30, 2018. https://www.washingtonpost .com/outlook/people-dont-vote-for-want-they-want-they-vote-for-who-they-are /2018/08/30/fb5b7e44-abd7-11e8-8a0c-70b618c98d3c_story.html.

Asch, Solomon. "Opinions and Social Pressure." *Scientific American* 193 (November 1955): 31–35.

Bardon, Adrian. *The Truth about Denial: Bias and Self-Deception in Science, Politics, and Religion*. Oxford: Oxford University Press, 2020.

Beck, Julie. "This Article Won't Change Your Mind." *Atlantic*, March 13, 2017.

Berman, Jonathan. *Anti-Vaxxers: How to Challenge a Misinformed Movement*. Cambridge, MA: MIT Press, 2020.

Berman, Mark. "More Than 100 Confirmed Cases of Measles in the U.S." *Washington Post*, February 2, 2015. https://www.washingtonpost.com/news/to-your-health/wp /2015/02/02/more-than-100-confirmed-cases-of-measles-in-the-u-s-cdc-says.

Boghossian, Peter, and James Lindsay. *How to Have Impossible Conversations: A Very Practical Guide*. New York: Lifelong Books, 2019.

Boseley, Sarah. "Mbeki Aids Denial 'Caused 300,000 Deaths.'" *Guardian*, November 26, 2008.

Branigin, Rose. "I Used to Be Opposed to Vaccines. This Is How I Changed My Mind." *Washington Post*, February 11, 2019. https://www.washingtonpost.com/opinions /i-used-to-be-opposed-to-vaccines-this-is-how-i-changed-my-mind/2019/02/11 /20fca654-2e24-11e9-86ab-5d02109aeb01_story.html.

Cassam, Quassim. *Conspiracy Theories*. Cambridge: Polity, 2019.

Clark, Daniel, dir. *Behind the Curve*. 2018; Delta-V Productions. https://www.behind thecurvefilm.com/.

Coll, Steve. *Private Empire: ExxonMobil and American Power*. New York: Penguin, 2012.

Cook, John. "A History of FLICC: The Five Techniques of Science Denial." *Skeptical Science*, March 31, 2020. https://skepticalscience.com/history-FLICC-5-techniques-science-denial.html.

Crease, Robert P. *The Workshop and the World: What Ten Thinkers Can Teach Us about the Authority of Science*. New York: Norton, 2019.

Dean, Cornelia. *Making Sense of Science: Separating Substance from Spin*. Cambridge, MA: Harvard University Press, 2017.

Deer, Brian. "British Doctor Who Kicked-Off Vaccines-Autism Scare May Have Lied, Newspaper Says." *Los Angeles Times*, February 9, 2009.

Deer, Brian. "How the Case against the MMR Vaccine Was Fixed." *British Medical Journal* 342 (2011): case 5347.

Diethelm, Pascal, and Martin McKee. "Denialism: What Is It and How Should Scientists Respond?" *European Journal of Public Health* 19, no. 1 (January 2009): 2–4. https://academic.oup.com/eurpub/article/19/1/2/463780.

Doyle, Alister. "Evidence for Man-Made Global Warming Hits Gold Standard." Reuters, February 25, 2019. https://www.reuters.com/article/us-climatechange-temperatures/evidence-for-man-made-global-warming-hits-gold-standard-scientists-idUSKCN1QE1ZU.

Festinger, Leon, Henry Ricken, and Stanley Schachter. *When Prophecy Fails*. New York: Harper and Row, 1964.

Folley, Aris. "NASA Chief Says He Changed Mind about Climate Change because He 'Read a Lot.'" *The Hill*, June 6, 2018. https://thehill.com/blogs/blog-briefing-room/news/391050-nasa-chief-on-changing-view-of-climate-change-i-heard-a-lot-of.

Foran, Clare. "Ted Cruz Turns Up the Heat on Climate Change." *Atlantic*, December 9, 2015.

Gee, David. "Almost All Flat Earthers Say YouTube Videos Convinced Them, Study Says." *Friendly Atheist*, February 20, 2019. https://friendlyatheist.patheos.com/2019/02/20/almost-all-flat-earthers-say-youtube-videos-convinced-them-study-says/.

Gillis, Justin. "Scientists Warn of Perilous Climate Shift within Decades, Not Centuries." *New York Times*, March 22, 2016.

Godlee, Fiona. "Wakefield Article Linking MMR Vaccine and Autism Was Fraudulent." *British Medical Journal* 342 (2011): case 7452.

Gorman, Sara, and Jack Gorman. *Denying to the Grave: Why We Ignore the Facts That Will Save Us*. Oxford: Oxford University Press, 2017.

Griswold, Eliza. "People in Coal Country Worry about the Climate, Too." *New York Times*, July 13, 2019. https://www.nytimes.com/2019/07/13/opinion/sunday/jobs -climate-green-new-deal.html.

Haidt, Jonathan. *The Righteous Mind: Why Good People Are Divided by Politics and Religion*. New York: Vintage, 2012.

Hall, Shannon. "Exxon Knew about Climate Change Almost 40 Years Ago." *Scientific American*, October 26, 2015.

Hamilton, Lawrence. "Conservative and Liberal Views of Science: Does Trust Depend on Topic?" *Carsey Research, Regional Issue Brief* 45 (Summer 2015). https://scholars .unh.edu/cgi/viewcontent.cgi?article=1251&context=carsey.

Hansen, James. *Storms of My Grandchildren*. New York: Bloomsbury, 2010.

Harris, Paul. "Four US States Considering Laws That Challenge Teaching of Evolution." *Guardian*, January 31, 2013.

Hoggan, James, and Richard Littlemore. *Climate Cover-Up: The Crusade to Deny Global Warming*. Vancouver: Greystone, 2009.

Hoofnagle, Mark. "About." *ScienceBlogs*, April 30, 2007. https://scienceblogs.com /denialism/about.

Huber, Rose. "Scientists Seen as Competent but Not Trusted by Americans." *Woodrow Wilson School* (September 22, 2014), https://publicaffairs.princeton.edu/news /scientists-seen-competent-not-trusted-americans.

Joyce, Christopher. "Rising Sea Levels Made This Republican Mayor a Climate Change Believer." *NPR*, May 17, 2016. https://www.npr.org/2016/05/17/477014145 /rising-seas-made-this-republican-mayor-a-climate-change-believer.

Kahan, Dan, E. Peters, E. Dawson, and P. Slovic. "Motivated Numeracy and Enlightened Self-Government." *Behavioural Public Policy*, preprint. Accessed October 25, 2020. https://pdfs.semanticscholar.org/2125/a9ade77f4d1143c4f5b15a534386e72e3aea.pdf.

Kahn, Brian. "No Pause in Global Warming." *Scientific American*, June 4, 2015.

Kahneman, Daniel. *Thinking Fast and Slow*. New York: Farrar, Straus and Giroux, 2011.

Kahn-Harris, Keith. *Denial: The Unspeakable Truth*. London: Notting Hill Editions, 2018.

Keeley, Brian. "Of Conspiracy Theories." *Journal of Philosophy* 96, no. 3 (March 1999): 109–126.

Kolbert, Elizabeth. "Why Facts Don't Change Our Minds." *New Yorker*, February 27, 2017. https://www.newyorker.com/magazine/2017/02/27/why-facts-dont-change-our -minds.

Krimsky, Sheldon. *GMOs Decoded: A Skeptic's View of Genetically Modified Foods*. Cambridge, MA: MIT Press, 2019.

Kruger, Justin, and David Dunning. "Unskilled and Unaware of It: How Difficulties in Recognizing One's Own Incompetence Lead to Inflated Self-Assessments." *Journal of Personality and Social Psychology* 77, no. 6 (1999): 1121–1134.

Kuklinski, James, et al. "Misinformation and the Currency of Democratic Citizenship." *Journal of Politics* 62, no. 3 (August 2000): 790–816.

Landrum, Asheley, Alex Olshansky, and Othello Richards. "Differential Susceptibility to Misleading Flat Earth Arguments on YouTube." *Media Psychology*, September 29, 2019. https://www.tandfonline.com/doi/full/10.1080/15213269.2019.1669461.

Leonard, Christopher. *Kochland: The Secret History of Koch Industries and Corporate Power in America*. New York: Simon and Schuster, 2019.

Lewandowky, Stephan, and John Cook. *The Conspiracy Theory Handbook*. 2020. https://www.climatechangecommunication.org/conspiracy-theory-handbook/.

Lewandowsky, Stephan, Gilles E. Gignac, and Klaus Oberauer. "Correction: The Role of Conspiracist Ideation and Worldviews in Predicting Rejection of Science." *PLoS ONE* 10, no. 8 (2015): e0134773.

Lewandowsky, Stephan, and Klaus Oberauer. "Motivated Rejection of Science." *Current Directions in Psychological Science* 25, no. 4 (2016): 217–222.

Lewandowsky, Stephan, Jan K. Woike, and Klaus Oberauer. "Genesis or Evolution of Gender Differences? Worldview-Based Dilemmas in the Processing of Scientific Information." *Journal of Cognition* 3, no. 1 (2020): 9.

Longino, Helen. *Science as Social Knowledge: Values and Objectivity in Scientific Inquiry*. Princeton: Princeton University Press, 1990.

Lynas, Mark. *Seeds of Science: Why We Got It So Wrong on GMOs*. London: Bloomsbury, 2018.

Lynch, Michael Patrick. *Know-It-All Society: Truth and Arrogance in Political Culture*. New York: Liveright, 2019.

Mason, Lilliana. "Ideologues without Issues: The Polarizing Consequences of Ideological Identities." *Public Opinion Quarterly* 82, no. S1 (March 21, 2018): 866–887.

Mayer, Jane. *Dark Money: The Hidden History of the Billionaires Behind the Rise of the Radical Right*. New York: Anchor, 2017.

McIntyre, Lee. "Flat Earthers, and the Rise of Science Denial in America." *Newsweek*, May 14, 2019. https://www.newsweek.com/flat-earth-science-denial-america-1421936.

McIntyre, Lee. "How to Talk to COVID-19 Deniers." *Newsweek*, August 18, 2020. https://www.newsweek.com/how-talk-covid-deniers-1525496.

McIntyre, Lee. *Post-Truth*. Cambridge, MA: MIT Press, 2018.

McIntyre, Lee. "The Price of Denialism." *New York Times*, November 7, 2015.

McIntyre, Lee. *Respecting Truth: Willful Ignorance in the Internet Age*. New York: Routledge, 2015.

McIntyre, Lee. *The Scientific Attitude: Defending Science from Denial, Fraud, and Pseudoscience*. Cambridge, MA: MIT Press, 2019.

Meikle, James, and Boseley, Sarah. "MMR Row Doctor Andrew Wakefield Struck Off Register." *Guardian*, May 24, 2010.

Mellor, D. H. "The Warrant of Induction." In *Matters of Metaphysics*. Cambridge: Cambridge University Press, 1991.

Mnookin, Seth. *The Panic Virus: The True Story Behind the Vaccine-Autism Controversy*. New York: Simon and Schuster, 2011.

Mooney, Chris. *The Republican Brain: The Science of Why They Deny Science—and Reality*. Hoboken, NJ: Wiley, 2012.

Mooney, Chris. *The Republican War on Science*. New York: Basic Books, 2005.

Moser, Laura. "Another Year, Another Anti-Evolution Bill in Oklahoma." *Slate*, January 25, 2016. http://www.slate.com/blogs/schooled/2016/01/25/oklahoma_evolution_controversy_two_new_bills_present_alternatives_to_evolution.html.

Nahigyan, Pierce. "Global Warming Never Stopped." *Huffington Post*, December 3, 2015, https://www.huffpost.com/entry/global-warming-never-stopped_b_8704128.

Nichols, Tom. *The Death of Expertise: The Campaign against Established Knowledge and Why It Matters*. Oxford: Oxford University Press, 2017.

NPR News. "Scientific Evidence Doesn't Support Global Warming, Sen. Ted Cruz Says." *NPR*, December 9, 2015. http://www.npr.org/2015/12/09/459026242/scientific-evidence-doesn-t-support-global-warming-sen-ted-cruz-says.

Nuccitelli, Dana. "Here's What Happens When You Try to Replicate Climate Contrarian Papers." *Guardian*, August 25, 2015.

Nyhan, Brendan, and Jason Reifler. "The Roles of Information Deficits and Identity Threat in the Prevalence of Misperceptions." *Journal of Elections, Public Opinion and Parties* 29, no. 2 (2019): 222–244.

Nyhan, Brendan, and Jason Reifler. "When Corrections Fail: The Persistence of Political Misperceptions." *Political Behavior* 32 (2010): 303–330.

O'Connor, Cailin, and James Weatherall. *The Misinformation Age: How False Beliefs Spread*. New Haven: Yale University Press, 2017.

Offit, Paul. *Deadly Choices: How the Anti-Vaccine Movement Threatens Us All*. New York: Basic Books, 2015.

Oreskes, Naomi, and Erik Conway. *Merchants of Doubts: How a Handful of Scientists Obscured the Truth on Issues from Tobacco Smoke to Global Warming*. New York: Bloomsbury, 2010.

Otto, Shawn. *The War on Science: Who's Waging It, Why It Matters, What We Can Do about It*. Minneapolis: Milkweed, 2016.

Pappas, Stephanie. "Climate Change Disbelief Rises in America" *LiveScience*, January 16, 2014. http://www.livescience.com/42633-climate-change-disbelief-rises.html.

Pigliucci, Massimo. *Denying Evolution: Creationism, Scientism and the Nature of Science*. Oxford: Sinauer Associates, 2002.

Pinker, Steven. *Enlightenment Now: The Case for Reason, Science, Humanism, and Progress*. New York: Penguin, 2019.

Plait, Phil. "Scientists Explain Why Ted Cruz Is Wrong about the Climate." *Mother Jones,* January 19, 2016.

Prothero, Donald. *Reality Check: How Science Deniers Threaten Our Future*. Bloomington: Indiana University Press, 2013.

Redlawsk, David, et al. "The Affective Tipping Point: Do Motivated Reasoners Ever 'Get It'?" *Political Psychology* 31, no. 4 (2010): 563–593.

Saslow, Eli. *Rising Out of Hatred: The Awakening of a Former White Nationalist*. New York: Anchor, 2018.

Schmid, Phillip, and Cornelia Betsch. "Effective Strategies for Rebutting Science Denialism in Public Discussions." *Nature Human Behaviour* 3 (2019): 931–939.

Shepphard, Kate. "Ted Cruz: 'Global Warming Alarmists Are the Equivalent of the Flat-Earthers.'" *Huffington Post*, March 25, 2015, https://www.huffpost.com/entry/ted-cruz-global-warming_n_6940188.

Shermer, Michael. *The Believing Brain*. New York: Times Books, 2011.

Shermer, Michael. "How to Convince Someone When Facts Fail: Why Worldview Threats Undermine Evidence." *Scientific American*, January 1, 2017. https://www.scientificamerican.com/article/how-to-convince-someone-when-facts-fail/.

Specter, Michael. *Denialism: How Irrational Thinking Hinders Scientific Progress, Harms the Planet, and Threatens Our Lives*. New York: Penguin, 2009.

Steinhauser, Jennifer. "Rising Public Health Risk Seen as More Parents Reject Vaccines." *New York Times*, March 21, 2008.

Storr, Will. *The Heretics: Adventures with the Enemies of Science*. New York: Picador, 2013.

Sun, Lena, and Maureen O'Hagan. "'It Will Take Off Like Wildfire': The Unique Dangers of the Washington State Measles Outbreak." *Washington Post*, February 6, 2019. https://www.washingtonpost.com/national/health-science/it-will-take-off-like -a-wildfire-the-unique-dangers-of-the-washington-state-measles-outbreak/2019/02 /06/cfd5088a-28fa-11e9-b011-d8500644dc98_story.html.

Trivers, Robert. *The Folly of Fools: The Logic of Deceit and Self-Deception in Human Life*. New York: Basic Books, 2011.

van der Linden, Sander. "Countering Science Denial." *Nature Human Behaviour* 3 (2019): 889–890. https://www.nature.com/articles/s41562-019-0631-5.

Warzel, Charlie. "How to Actually Talk to Anti-Maskers." *New York Times*, July 22, 2020. https://www.nytimes.com/2020/07/22/opinion/coronavirus-health-experts.html.

West, Mick. *Escaping the Rabbit Hole: How to Debunk Conspiracy Theories Using Facts, Logic, and Respect*. New York: Skyhorse, 2018.

Wood, Thomas, and Ethan Porter. "The Elusive Backfire Effect." August 5, 2016. https://djflynn.org/wp-content/uploads/2016/08/elusive-backfire-effect-wood -porter.pdf.

Zimring, James. *What Science Is and How It Really Works*. Cambridge: Cambridge University Press, 2019.

Index